Springer Theses

Recognizing Outstanding Ph.D. Research

Aims and Scope

The series "Springer Theses" brings together a selection of the very best Ph.D. theses from around the world and across the physical sciences. Nominated and endorsed by two recognized specialists, each published volume has been selected for its scientific excellence and the high impact of its contents for the pertinent field of research. For greater accessibility to non-specialists, the published versions include an extended introduction, as well as a foreword by the student's supervisor explaining the special relevance of the work for the field. As a whole, the series will provide a valuable resource both for newcomers to the research fields described, and for other scientists seeking detailed background information on special questions. Finally, it provides an accredited documentation of the valuable contributions made by today's younger generation of scientists.

Theses may be nominated for publication in this series by heads of department at internationally leading universities or institutes and should fulfill all of the following criteria

- They must be written in good English.
- The topic should fall within the confines of Chemistry, Physics, Earth Sciences, Engineering and related interdisciplinary fields such as Materials, Nanoscience, Chemical Engineering, Complex Systems and Biophysics.
- The work reported in the thesis must represent a significant scientific advance.
- If the thesis includes previously published material, permission to reproduce this must be gained from the respective copyright holder (a maximum 30% of the thesis should be a verbatim reproduction from the author's previous publications).
- They must have been examined and passed during the 12 months prior to nomination.
- Each thesis should include a foreword by the supervisor outlining the significance of its content.
- The theses should have a clearly defined structure including an introduction accessible to new PhD students and scientists not expert in the relevant field.

Indexed by zbMATH.

More information about this series at http://www.springer.com/series/8790

Cédric Schmidt

Time-Resolved Soft X-Ray Absorption Spectroscopy of Molecules in the Gas and Liquid Phases

Doctoral Thesis accepted by
University of Geneva, Geneva, Switzerland

Springer

Author
Dr. Cédric Schmidt
Department of Applied Physics
University of Geneva
Geneva, Switzerland

Supervisor
Prof. Jean-Pierre Wolf
Department of Applied Physics
University of Geneva
Geneva, Switzerland

ISSN 2190-5053 ISSN 2190-5061 (electronic)
Springer Theses
ISBN 978-3-030-67837-1 ISBN 978-3-030-67838-8 (eBook)
https://doi.org/10.1007/978-3-030-67838-8

This Springer imprint is published by the registered company Springer Nature Switzerland AG
The registered company address is: Gewerbestrasse 11, 6330 Cham, Switzerland

Supervisor's Foreword

The work of Cédric Schmidt is dedicated to the study of the relaxation dynamics of molecules in both the gas and liquid phases after strong-field ionization, using transient absorption in the soft X-rays. This work has been performed in close collaboration with the group of Prof. H. J. Wörner at ETHZ, with whom we have been collaborating for many years within the Swiss NCCR MUST network.

In particular, this Ph.D. work presents the first realization of time-resolved X-ray absorption spectroscopy (TR-XAS) in the spectral water window with a laser-based HHG source. These remarkable experiments were not only performed for isolated molecules, but also in liquids, for which the spectral coverage of the K-edges of C, N, and O are of primary importance to investigate biological molecules.

In Chap. 1, C. Schmidt provides the theoretical framework required to analyze the experimental results. He introduces the basic concepts of high field laser-matter interactions, leading to filamentation, HHG, strong-field ionization, propagation effects, and soft X-ray (SXR) absorption spectroscopy.

Chapter 2 is dedicated to the detailed description of the experimental setup, from the laser-OPA (centered at 2 μm) source optimization, including Carrier-Envelope-source (CEP) stabilization, to the different vacuum chambers, including the HHG generation and target chambers, the liquid jet that was used, and the XUV-SXR spectrometer. A very nice analysis of the spectral characteristics of the SXR spectrometer and its optimization in terms of sensitivity and resolution (better than 1 eV) over the water window is presented. The chapter also shows the optimization of the HHG source as a function of pressure, gas target, and laser intensity and wavelength.

Chapter 3 describes the results of the TR-XAS measurements of the relaxation dynamics of strong-field ionized molecules in the gas phase. These remarkable results highlight the first time-resolved measurements in the soft X-ray water window using a table-top HHG source instead of a large facility. More precisely, the experiment allowed previously unexamined light-induced chemical reactions in the lowest electronic states of isolated CF_4^+ and SF_6^+ molecules in the gas phase to be followed. By probing element-specific core-to-valence transitions at the carbon K-edge or the sulfur L-edges (spectral region: 150 eV-350 eV), it was possible to characterize their reaction paths (bond breaking of a fluorine atom) and observe the effect of symmetry-breaking through the splitting of absorption bands and Rydberg-valence mixing induced by the geometry changes.

Following these excellent results, C. Schmidt decided to develop a double stage filamentation setup to reduce the pulse duration of the 2 μm laser pulses and increase its peak power. In his Chap. 4, he describes this development in detail, which allowed him to obtain pulses as short as 1.5 cycle, and peak powers of 0.2 TW. With this very performant infrared driver, C. Schmidt extended the HHG cut-off in Helium beyond the oxygen K-edge at 540 eV, providing a SXR spectrum for TR-XAS experiments that cover the whole water window.

Chapter 5 concentrates on the use of this extended SXR source for TR-XAS of molecules in the liquid phase. To this end, a flat liquid jet, developed and constructed at the ETHZ by H. J. Wörner's group was implemented in the experiment in Geneva. Very exciting experimental results are described in this chapter for alcohols (at both the carbon and oxygen K-edges) and pyridine (at both the C and the N K-edges). The results on pyridine are particularly impressive, showing clear early dynamics in the 50 fs timescale and longer timescale modulations related to nuclear re-arrangement.

In conclusion, the work performed by C. Schmidt in his TR-XAS study of molecules in the gas and liquid phases in the water window is truly remarkable. It is, therefore, a real pleasure for me to write this foreword.

Geneva, Switzerland
January 2021

Prof. Jean-Pierre Wolf

Abstract

My thesis relates to the study of ultrafast molecular dynamics through the development of a soft X-ray laser source. The transient absorption experiment described in this manuscript combines the latest generation of femtosecond lasers with a vacuum spectroscopy system enabling time-resolved measurements of samples at the low femtosecond timescale with a transmission geometry. The technique relies on the generation of high order harmonics (HHG). Using the atomic selectivity of high energies and the temporal coherence of laser technology, we demonstrate the observation of the first stages of chemical transformation of matter in the gas and liquid phases.

Contents

Acronyms

ATI	Above threshold ionization
BBO	Beta barium borate, β-BaB_2O_4
BiBO	Bismuth borate, BiB_3O_6
CCD	Charge-coupled device
CEP	Carrier-envelope phase
CPA	Chirped-pulse amplification
CXRO	Center for X-ray optics
DFG	Difference-frequency generation
DFT	Density functional theory
EXAFS	Extended X-ray absorption fine structure
FEL	Free electron laser
FROG	Frequency-resolved optical gating
FS	Fused silica
FWHM	Full width at half maximum
GDD	Group delay dispersion
GVD	Group velocity dispersion
HHG	High harmonic generation / generated
IP	Ionization potential
IR	Infrared (700 nm - 1 mm)
LASER	Light amplification by stimulated emission of radiation
MCP	Micro-channel plate
NIR	Near-infrared (0.75-1.4 μm)
OPA	Optical parametric amplifier / amplification
OPCPA	Optical parametric chirped pulse amplification
SHG	Second harmonic generation
SPM	Self-phase modulation
SWIR	Short-wavelength infrared (1.4-3 μm)
SXR	Soft X-ray (0.12-10 nm, 124-10 000 eV)
TDDFT	Time-dependent density functional theory
THG	Third harmonic generation

TOD	Third order dispersion
TR	Time-resolved
UV	Ultraviolet (100-400 nm, 3.10-12.4 eV)
XANES	X-ray absorption near edge structure
XAS	X-ray absorption spectroscopy
XUV	Extreme ultraviolet (10-124 nm, 10-124 eV)

Chapter 1
Introduction

The development of science over the course of history has shown that certain physical phenomena occur on extremely long scales with respect to the human time. The genesis of the Milky Way, for example, resulted from the violent fusion of protogalaxies and it took a few billion years (10^{18} s) of the effect of the gravitational force before stars were formed in the center of its giant dark matter halo [1]. In contrast, there are phenomena that occur so quickly that they are beyond our ability to decipher them with present technology. Charge migration in a molecule, for instance, can occur on a time scale of a few hundred attoseconds (10^{-16} s) [2]. At these time scales, the entities at play must be regarded as wave functions, and their evolution should be described by an appropriate quantum theory, such as the Schrödinger equation.

Relying on the vast discoveries in quantum physics over the last century, a detailed understanding of the dynamics of the building blocks in nature is now becoming accessible. Specifically, getting more insight on electron transport mechanisms could also bring technological benefits: the rearrangement of the electron density during structural changes and how it relates to theoretical models might contribute to the development of alternative energy sources, complex drug synthesis, and novel solutions for electronics in the post-Moore's law era.

In this perspective, robust new experimental measurements are needed to confirm existing theoretical models and provide feedback and input values necessary to strengthen the validity and predictive power of numerical simulations. Experiments typically rely on ever-shorter sources of electromagnetic radiation from the infrared (IR) to bursts of X-rays, electron diffraction techniques and recent solutions developed in the field of magnetic resonance.

X-ray spectroscopy and its application to the study of structural properties of matter is one of the greatest successes of science in the twentieth century. Over the last few years, pioneering works have granted access to the temporal dimension to synchrotrons and X-ray free-electron lasers (XFELs) [3, 4]. Indeed, the emergence of XFEL light sources all around the world in the late 2000s (LCLS in Stanford started in 2009, SACLA in Japan, PAL in Korea, SwissFEL in Switzerland and EuXFEL in

C. Schmidt, *Time-Resolved Soft X-Ray Absorption Spectroscopy of Molecules in the Gas and Liquid Phases*, Springer Theses,
https://doi.org/10.1007/978-3-030-67838-8_1

Germany) opened new opportunities for the analysis of submicrometric crystals and for structural biology imaging with atomic resolution at the femtosecond timescale. Aside from these large facilities, incoherent tabletop hard X-ray sources have also been applied to picosecond time-resolved studies [5, 6].

Another great success of optical science in the mid twentieth century is the invention of the laser (Light Amplification through Stimulated Emission of Radiation), a device based on the process of stimulated emission first predicted by A. Einstein in 1916 [7]. The investigation of laser-matter interaction started right after the first experimental demonstration of solid-state laser emission in a ruby crystal by T.H. Maiman in 1960 [8]. Shortly, the fast development of large emission bandwidth oscillators based on neodymium-doped yttrium aluminum garnet crystals (Nd:YAG) or on titanium-doped sapphire crystals (Ti:Sa) enabled producing ultrashort pulses through Kerr-lens mode locking [9, 10]. This opened broad new possibilities for laser research and applications by accessing unprecedented time resolutions and electric-field intensities [11].

Following these progresses, the high-harmonic generation (HHG) process was observed in gases for the first time in 1987 by research groups in France and in the United States while focusing a picosecond laser into a gas target [12, 13]. Later initiated by numerical experiments [14], the explanation of the generation mechanism in the nonperturbative regime was introduced by P. Corkum in 1993 [15]. Successively, the analytical approach provided the physical picture of HHG as a three-step process and predicted the generation of attosecond pulses [16]. Therefore, strong efforts were aimed at temporally characterizing and isolating the generated attosecond pulses since the first years after the discovery of HHG [17]. At the same time, the extension of the photon energy cut-off into the X-ray regime [18, 19] and the first demonstration of a frequency comb in the vacuum ultraviolet energies enabled by HHG [20] mark key innovations for HHG source development. While the first high-order harmonics and attosecond pulses have been generated in a gas jet, other approaches point towards new generation schemes, including solid-state based HHG [21], plasma surfaces [22], gas-filled capillaries [23] and generation in liquids [24].

While pioneering HHG based time-resolved pump-probe experiments were performed using near-infrared (NIR) laser-based sources with a cut-off limited to the extreme ultraviolet energy range [25–27], most spectroscopic studies providing both sub-femtosecond time resolution and atomic spatial resolution require core electron excitation at significantly higher energies. In particular, the dynamics of organic and biological molecules would be optimally addressed in the so-called water transparency window, which lies between the carbon and oxygen K-shell absorption edges (284–543 eV) enabling X-ray absorption measurements in aqueous solutions [28, 29].

Along with the progresses in optical parametric amplified laser sources [30], sufficient power and repetition rate are now available in the short-wavelength infrared (SWIR) range enabling the generation of soft X-ray (SXR) pulses covering the water window [29, 31–36]. Moreover, techniques such as attosecond streaking [37–39] have recently been extended to these energies, enabling the generation of the shortest attosecond pulses ever produced [40]. These studies constantly move forward the field by increasing the photon flux and by attempting to observe light induced ultrafast molecular dynamics with the outstanding time-resolution provided by HHG. All the efforts described in this manuscript go in the same direction by applying original ideas.

Although achieving good SXR fluxes at the high energy side (500 eV) of the spectrum remains challenging in reason of the extremely nonlinear process of HHG requiring available laser pulses to be further optimized to achieve higher pulse intensities. Such intensities are usually obtained by reducing pulse duration via spectral broadening in gas-filled hollow fibers [41–45] or by filamentation [46–49]. The later approach has received increasing attention as it enables a cost-effective and robust method for intense laser pulse compression. Indeed, producing few-cycle SWIR pulses via filamentation of femtosecond laser pulses in gases has been described in numerous studies after the introduction of tabletop amplified laser sources capable of reaching the critical power for Kerr self-focusing [46–49]. Some striking results reported highly efficient pulse compression schemes reaching less than 10 fs in the NIR [47].

This thesis unveils the achievement of the first realization of time-resolved X-ray absorption spectroscopy (TR-XAS) in the water window with a HHG source [50]. Developed in close collaboration with the Ultrafast Spectroscopy and Attosecond Science group of H.J. Wörner in ETH Zürich, this tabletop experiment on the observation of molecular dissociation in the gas phase, overcomes many of the challenges faced by researchers at large facilities. The technical developments realized in order to fully cover the water window spectral range are presented and some preliminary results of the extension of this approach to the liquid phase are provided.

The first chapter provides the theoretical framework for interpreting the experimental results and the technical developments relative to the laser source. It introduces the basics of molecular structures and the effects of perturbations induced by a strong laser field. The different laser-matter interaction regimes are presented together with the principles of intense laser pulse propagation, HHG and SXR absorption spectroscopy.

The second chapter presents the experimental setup in detail together with the data acquisition and analysis methods, with a special emphasis on the transient absorption procedure. The different laser pulses involved in the measurements are characterized and the optical optimizations aiming to increase the system performances are detailed. The HHG conditions for SXR generation are discussed and compared to equivalent systems performing in the water window. Finally, the design and the modifications applied to the detection spectrometer are presented together with suggestions for further improvements.

The third chapter focuses on the results of TR-XAS measurements of molecular dissociation in the gas phase at the low energy side of the water-window [50]. Using the transient absorption technique, strong-field induced dissociation of carbon tetrafluoride CF_4 and sulfur hexafluoride SF_6 was observed with a femtoseconds time resolution. Details of the quantum calculations of core-level transitions realized by theoreticians at ETH Zürich are provided together with the spectral assignments of the time-resolved structures. The descriptions and analysis reviewed in this chapter follow and complement the published physical interpretation of the observed molecular transformations [50].

The fourth chapter presents the results of the extension of the source beyond the oxygen K-edge energy [31]. The adopted two-stage filamentation scheme for pulse broadening coupled to a post-compression technique based on transmission through a substrate is discussed and the resulting sub-two-cycle SWIR pulse is spatially and temporally characterized. Generated in helium, the highest ionization energy gas, the resulting SXR spectrum fully covers the water window energy range with similar flux conditions as reached for the successful gas dissociation measurements presented in Chap. 3.

Finally, the fifth chapter provides preliminary experimental observations of liquid pyridine C_5H_5N X-ray absorption near-edge structure (XANES) modifications following short field NIR interaction. By coupling a vacuum flat jet interaction chamber with the HHG water window source described in Chap. 2 and taking advantage of the cut-off extension presented in Chap. 4, the induced reaction is tracked simultaneously from the carbon and nitrogen K-edges at sub-100 fs time resolution.

References

1. Garry and Feurer T (2016) A journey into time in powers of ten. vdf Hochschulverlag
2. Kraus PM et al (2015) Measurement and laser control of attosecond charge migration in ionized iodoacetylene. Science 350(6262):790–795
3. Bressler C et al (2009) Femtosecond XANES study of the light-induced spin crossover dynamics in an iron(II) complex. Science 323(5913):489–492
4. Wernet P et al (2015) Orbital-specific mapping of the ligand exchange dynamics of Fe(CO)5 in solution. Nature 520(7545):78–81
5. Ráksi F, Wilson KR, Jiang Z, Ikhlef A, Côté CY, Kieffer J (1996) Ultrafast x-ray absorption probing of a chemical reaction. J Chem Phys 104(15):6066–6069
6. Bargheer M et al (2004) Coherent atomic motions in a nanostructure studied by femtosecond X-ray diffraction. Science 306(5702):1771–1773
7. Einstein A (1916) Strahlungs-emission und absorption nach der Quantentheorie. Dtsch Phys Gesellschaft 18:318–323
8. Maiman TH (1960) Stimulated optical radiation in ruby. Nature 187(4736):493–494
9. Boyd RW (2008) Nonlinear optics, 3rd Edition
10. Cerullo G, De Silvestri S, Magni V (1994) Self-starting Kerr-lens mode locking of a Ti:sapphire laser. Opt Lett 19(14):1040
11. Svelto O, Hanna DC (1998) Principles of laser. Plenum Press
12. McPherson A et al (1987) Studies of multiphoton production of vacuum-ultraviolet radiation in the rare gases. J Opt Soc Am B 4(4):595
13. Ferray M, L'Huillier A, Li XF, Lompre LA, Mainfray G, Manus C (1988) Multiple-harmonic conversion of 1064 nm radiation in rare gases. J Phys B at Mol Opt Phys 21(3):L31–L35
14. Krause JL (1992) High-order harmonic generation from atoms and ions in the high intensity regime. Phys Rev Lett 68(24):3535–3538
15. Corkum PB (1993) Plasma perspective on strong field multiphoton ionization. Phys Rev Lett 71(13):1994–1997
16. Lewenstein M, Balcou P, Ivanov MY, L'Huillier A, Corkum PB (1994) Theory of high-harmonic generation by low-frequency laser fields. Phys Rev A 49(3):2117–2132
17. Hentschel M et al (2001) Attosecond metrology. Nature 414(6863):509–513

18. Popmintchev T et al (2012) Bright coherent ultrahigh harmonics in the keV X-ray regime from mid-infrared femtosecond lasers. Science 336(6086):1287–1291
19. Seres E, Seres J, Krausz F, Spielmann C (2004) Generation of coherent soft-X-ray radiation extending far beyond the titanium L edge. Phys Rev Lett 92(16):163002
20. Goble C et al (2005) A frequency comb in the extreme ultraviolet. In: Digest of the LEOS summer topical meetings, pp 59–60
21. Ghimire S, Dichiara AD, Sistrunk E, Agostini P, Dimauro LF, Reis DA (2011) Observation of high-order harmonic generation in a bulk crystal. Nat Phys 7(2):138–141
22. Tsakiris GD, Eidmann K, Meyer-ter-Vehn J, Krausz F (2006) Route to intense single attosecond pulses. New J Phys 8:19–19
23. Rundquist A, Chang Z, Herne C, Backus S, Murnane MM, Kapteyn HC (1998) Phase-matched generation of coherent soft X-rays. Science 280(29):1412–1415
24. Luu TT et al (2018) Extreme–ultraviolet high–harmonic generation in liquids. Nat Commun 9(1):3723
25. Goulielmakis E et al (2010) Real-time observation of valence electron motion. Nature 466(7307):739–743
26. Holler M, Schapper F, Gallmann L, Keller U (2011) Attosecond electron wave-packet interference observed by transient absorption. Phys Rev Lett 106(12):123601
27. Schultze M et al (2014) Attosecond band-gap dynamics in silicon. Science 346(6215):1348–1352
28. Jordan I, Huppert M, Brown MA, van Bokhoven JA, Wörner HJ (2015) Photoelectron spectrometer for attosecond spectroscopy of liquids and gases. Rev Sci Instrum 86(12):123905
29. Kleine C et al (2019) Soft X-ray absorption spectroscopy of aqueous solutions using a table-top femtosecond soft X-ray source. J Phys Chem Lett 10(1):52–58
30. Cerullo G, De Silvestri S (2003) Ultrafast optical parametric amplifiers. Rev Sci Instrum 74(1):1–18
31. Schmidt C et al (2018) High-order harmonic source spanning up to the oxygen K-edge based on filamentation pulse compression. Opt Express 26(9):11834
32. Teichmann SM, Silva F, Cousin SL, Hemmer M, Biegert J (2016) 0.5-keV soft X-ray attosecond continua. Nat Commun 7:11493
33. Johnson AS et al (2018) High-flux soft x-ray harmonic generation from ionization-shaped few-cycle laser pulses. Sci Adv 4(5):eaar3761
34. Attar AR et al (2017) Femtosecond x-ray spectroscopy of an electrocyclic ring-opening reaction. Science 356(6333):54–59
35. Popmintchev D et al (2018) Near- and extended-edge X-ray-absorption fine-structure spectroscopy using ultrafast coherent high-order harmonic supercontinua. Phys Rev Lett 120(9):1–6
36. Buades B et al (2018) Attosecond-resolved petahertz carrier motion in semi-metallic TiS_2. arXiv:1808.06493
37. Saito N, Ishii N, Kanai T, Watanabe S, Itatani J (2016) Attosecond streaking measurement of extreme ultraviolet pulses using a long-wavelength electric field. Sci Rep 6(1):35594
38. Cousin SL et al (2017) Attosecond streaking in the water window: a new regime of attosecond pulse characterization. arXiv:1710.02208
39. Li J et al (2017) 53-attosecond X-ray pulses reach the carbon K-edge. Nat Commun 8(1):186
40. Gaumnitz T et al (2017) Streaking of 43-attosecond soft-X-ray pulses generated by a passively CEP-stable mid-infrared driver. Opt Express 25(22):27506–27518
41. Suda A, Hatayama M, Nagasaka K, Midorikawa K (2003) Generation of 9-fs 5-mJ pulses using a hollow fiber with pressure gradient. In: OSA trends in optics and photonics series
42. Schmidt BE et al (2010) Compression of 1.8 μm laser pulses to sub two optical cycles with bulk material. Appl Phys Lett 96(12):121109
43. Béjot P, Schmidt BE, Kasparian J, Wolf JP, Legaré F (2010) Mechanism of hollow-core-fiber infrared-supercontinuum compression with bulk material. Phys Rev A At Mol Opt Phys 81(6)
44. Silva F, Teichmann SM, Cousin SL, Hemmer M, Biegert J (2015) Spatiotemporal isolation of attosecond soft X-ray pulses in the water window. Nat Commun 6:6611

45. Cousin SL, Silva F, Teichmann S, Hemmer M, Buades B, Biegert J (2014) High-flux table-top soft x-ray source driven by sub-2-cycle, CEP stable, 1.85-μm 1-kHz pulses for carbon K-edge spectroscopy. Opt Lett 39(18):5383–5386
46. Hauri CP et al (2004) Generation of intense, carrier-envelope phase-locked few-cycle laser pulses through filamentation. Appl Phys B 79(6):673–677
47. Stibenz G, Zhavoronkov N, Steinmeyer G (2006) Self-compression of millijoule pulses to 7.8 fs duration in a white-light filament. In: 2006 conference on lasers and electro-optics and 2006 quantum electronics and laser science conference, pp. 1–2
48. Skupin S et al (2006) Self-compression by femtosecond pulse filamentation: experiments versus numerical simulations. Phys Rev E Stat Nonlinear, Soft Matter Phys
49. Couairon A, Mysyrowicz A (2007) Femtosecond filamentation in transparent media. Phys Rep 441(2):47–189
50. Pertot Y et al (2017) Time-resolved x-ray absorption spectroscopy with a water window high-harmonic source. Science 355(6322):264–267

Chapter 2
Theoretical Background

The purpose of this chapter is to provide the theoretical framework for interpreting the experiments presented in this thesis. It aims to understand the dynamic responses of molecules in the gas and liquid phases induced by a strong laser field. This is a rather challenging prospect. Indeed, the intricacy of molecular systems, which requires multiple mechanisms to interplay for a specific outcome, usually complicates the interpretation of experimental results. If quantitative numerical simulations help sorting out the multiplicity of parameters, the strong field interaction, however, generally considerably distorts the molecular system and adds to its complexity. The corresponding quantum calculations can become extremely challenging, requiring deep computational skills and knowledge of theory to be reliable. For the purpose of this thesis, various interaction regimes were implemented, both for the optimization of setup performances and for the molecular dynamics investigation. This chapter thus offers an overview of these regimes together with the basics of intense laser pulse propagation, high-harmonic generation and soft X-ray absorption spectroscopy.

2.1 Concepts of Molecular Orbital Theory

The molecular orbital theory is one of the two main frameworks that describe the electronic structure of molecules using quantum mechanics. Proposed in the early twentieth century, it complements the valence bond theory by treating the electrons as moving under the influence of the nuclei in the whole molecular system instead of being associated with individual bonds between atoms [1]. The energetic and spatial characteristics of a bond electron is described as a molecular orbital surrounding two or more atoms in a molecule. These orbital states are approximated as linear combinations of atomic orbitals (LCAO). Such approximations can be calculated using the density functional theory (DFT) or the Hartree–Fock models to solve the Schrödinger equation.

C. Schmidt, *Time-Resolved Soft X-Ray Absorption Spectroscopy of Molecules in the Gas and Liquid Phases*, Springer Theses,
https://doi.org/10.1007/978-3-030-67838-8_2

2.1.1 The Linear Combinations of Atomic Orbitals (LCAO) Method

In the LCAO approach, and in order to simplify the calculations for complex systems, it is assumed that only the electrons of an outer shell can contribute to the bond formation, whereas the ones of the core shells, i.e. close to the nucleus, remain unperturbed [2]. Therefore, for short distances in respect of the nuclei, the molecular orbital converges to the corresponding atomic orbital. As the time-independent Schrödinger equation is linear, the molecular orbitals can be approximated as LCAO. Assuming that the atomic wave functions $\{\psi_{\mathrm{A,i}}\}$ form a basis of the molecular Hamiltonian, their linear combination is considered as a trial solution for the molecular wave function [3]:

$$\Psi_{\mathrm{LCAO}} = \sum_i c_i \psi_{\mathrm{A,i}} \tag{2.1}$$

where the c_i coefficients can be solved numerically by substituting this equation into the Schrödinger equation and applying the variational principle.

2.1.2 Classification of Orbital States

Depending on the linear combinations, the orbitals can first be divided into three types. In a "bonding" orbital (symmetric combination), the electron density is concentrated in the region between a given pair of atoms and tends to attract each of the two nuclei toward the other in order to hold the two atoms together [4]. An "anti-bonding" orbital (antisymmetric combination) typically weakens the bond between two nuclei by concentrating the electron density in the shadow of each nucleus. Electrons in "non-bonding" orbitals tend to be associated with atomic orbitals that do not interact positively nor negatively with a second atom and that does not contribute to distort any bond strength [4].

Each orbital can be further classified by the symmetry properties of the associated electron density. Such symmetry consideration is widely used in spectroscopy to label the energy levels. For instance, a diatomic molecule has 3 symmetry properties: the molecular axis, the molecular central point and a plane containing the molecular axis. All these transformations leave the molecular Hamiltonian invariant, in such a way that they represent a full set of simultaneous eigenfunctions for the molecular system. This means that their quantum numbers uniquely identify each possible electronic energy level [3].

The rotational symmetry along the molecular axis is associated with the conservation of the projection of angular momentum L_z on the molecular axis. The related eigenvalue equation with the quantum number Λ is given by [5]:

$$L_z\psi_e = \pm\Lambda\hbar\psi_e \quad \Lambda = 0, 1, 2, \ldots \tag{2.2}$$

where the state labeling involves Greek letters such that $\Lambda = 0, 1, 2, \cdots \leftrightarrow \Sigma, \Pi, \Delta, \ldots$. This notation is analog to the notation of atomic orbitals, in which lower-case Greek letters are used for the angular momentum quantum number projection. This notation is also used to denote single electron states within a molecular system [3].

The central inversion symmetry is related to the parity of the electronic wave function. If its sign changes (does not change) upon inversion, the parity is odd (even). The associated spectroscopic notation is the subscript u (g).

For the planar symmetries, the sign upon reflexion is given as a superscript. However, the classification in terms of projection of the electronic angular momentum is no more relevant because of the broken cylindrical symmetry of the molecule and its electronic wave functions.

Finally, depending on its electron pairing, a quantum state can have different degree of degeneracy. The degeneracy level is written as the multiplicity $2S + 1$ with S the total spin of the molecular eigenstate. The molecular term symbol of an electronic state for a diatomic molecule employ symmetry labels from group theory and has the following form: ${}^{2S+1}\Lambda^{+/-}_{\Omega,g/u}$ where Ω is the projection of the total angular momentum along the internuclear axis. For heteronuclear diatomics, the term symbol does not include the g/u subscript, as there is not inversion center in the molecule. Polyatomic molecules follow the same labeling principle. For a given symmetry, the equivalent labeling can be found in the table for group theory [5].

2.1.3 *The Molecular Orbital Diagram*

The molecular orbital energy levels of a molecule can be illustrated in a molecular orbital diagram as shown in Fig. 2.1. The horizontal lines in the center represent the molecular orbitals and they are linked to their atomic orbital constituents for comparison. The energy levels typically increase from the bottom to the top and degenerated energy levels are commonly shown side by side. Appropriate orbital levels are filled by electrons respecting the Pauli exclusion principle. Every electron is symbolized by a small arrow accounting for its spin. For simple polyatomic molecules with a "central atom" such as tetrafluoromethane (CF_4), the diagram shows one of the identical bonds to the central atom. For more complex polyatomic molecules, the diagram can only show one bond of interest, leaving others out for simplicity. The low-level atomic orbitals and inner molecular orbitals are generally omitted from the diagram.

Once the system is filled with all the available electrons, the highest state containing at least one electron is referred to as the highest occupied molecular orbital (HOMO), where the lowest empty state is called the lowest unoccupied molecular

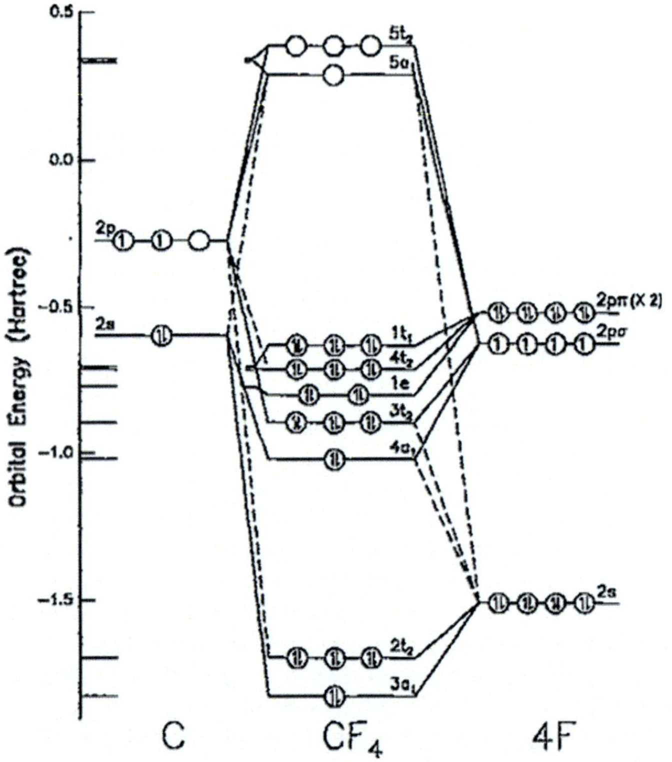

Fig. 2.1 Molecular orbital diagrams for the valence shells of CF_4 derived from a self-consistent-field calculation with a 6-31 G basis set. Solid and dashed lines are used to indicate strong and weaker atomic components for a particular molecular orbital, respectively. Reprinted from [6], with the permission of AIP Publishing

orbital (LUMO). In the case of CF_4, the HOMO is the $1t_1$ molecular orbital whereas the LUMO is the $5a_1$ orbital.

2.1.4 Radiative Transitions

When its electronic charge distribution interacts with a laser electric field, the molecule can absorb energy and undergo a transition from its initial energy state Ψ_i to a final state Ψ_f. Such radiative transition is possible only if it satisfies quantum mechanical selection rules related to angular momentum conservation or symmetry arguments. By neglecting the rotational energy of the molecule, and assuming the dipole approximation and the Born–Oppenheimer approximation [7], the total wave function can be written $\Psi = \psi\nu$, separating the electronic ψ and vibrational ν contributions. In the same way, the transition dipole moment $\mu = \mu_e + \mu_N$ between the initial and final states consists of an electronic part μ_e and a nuclear part μ_N leading to the projection:

$$\langle \Psi_f | \mu | \Psi_i \rangle = \langle \psi_f | \psi_i \rangle \langle \nu_f | \mu_N | \nu_i \rangle + \langle \nu_f | \nu_i \rangle \langle \psi_f | \mu_e | \psi_i \rangle \tag{2.3}$$

In the case of an electronic transition, $\psi_f \neq \psi_i$ and because of their orthogonality: $\langle \psi_f | \psi_i \rangle = 0$. Therefore, the transition probability $M_{i \to f}$ goes like [8]:

$$M_{i \to f} \propto |\langle \nu_f | \nu_i \rangle|^2 \times |\langle \psi_f | \mu_e | \psi_i \rangle|^2 \tag{2.4}$$

and thus, depends on the dipole transition probability between the two electronic states and on the overlap between initial and final vibrational states $|\langle \nu_f | \nu_i \rangle|^2$, called the Franck–Condon factor. When using the Born–Oppenheimer approximation, it is assumed that the transition takes place in a quasistatic regime. The internuclear distances are therefore considered constants, and the transition is referred to as a vertical transition. The corresponding projection region on Ψ_f is called the Franck–Condon region [9, 10].

2.1.5 Dissociation Mechanisms

It is possible to count multiple purely laser-induced dissociation pathways that may occur in a molecule. A radiative transition, as described in Sect. 2.1.4, may lead to a dissociative state or a vibrational state above the dissociation limit. Furthermore, the potential energy surface of the molecule can be dressed and deformed in such a way that dissociation is either assisted, reduced or delayed in time [11].

Figure 2.2 shows the three possible pathways for photodissociation in the perturbative regime. (1) shows an electronic transition to a repulsive state. Due to the Coulombic repulsion, these curves are very steep for small internuclear distances. The dissociation fragments typically acquire large kinetic energies E_{kin} as illustrated in the figure. This dissociation mechanism is called Coulomb explosion [12]. (2) represents an electronic transition to a vibrational state. As it lies above the dissociation limit, it is called an unbound state and the excess energy above the potential barrier is usually on the order of the vibrational level spacing. Finally, (3) illustrates a predissociation process, in which a transition to a bound excited state initially takes place. If this excited state is coupled to a dissociative state on another potential energy surface, the molecule can dissociate. Such couplings typically rely on internal spin–orbit effects and conical intersections or external collisions [13].

In the strong field regime, transitions to dissociative states can also take place through tunnel ionization as presented in Sect. 2.2.3.

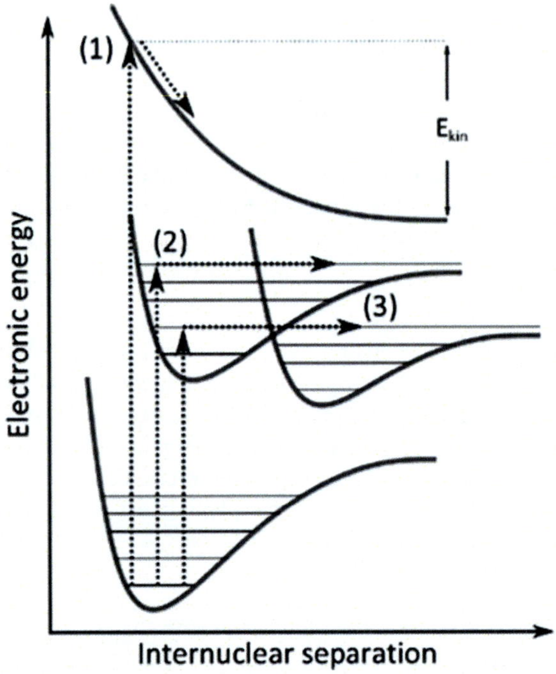

Fig. 2.2 Dissociation pathways in a molecular system: (1) transition to a repulsive state, (2) transition to an unbound state above the dissociation limit, (3) transition to a bound excited state coupled to a dissociative state by conical intersection. Reprinted from [11] with the kind permission of the author

2.2 Ionization

The interraction between an electric field and a molecular (atomic) system can lead to the release of electrons so that the remaining system, the ion, is positively charged and the free electrons are accumulated in the continuum. Starting from the current density equation $\vec{J} = q_e \rho \vec{v}_e$ (with q_e the elementary charge, $\vec{v}_e$ the velocity of the electrons and ρ the electron density), the divergence theorem and the Newton's second law, the differential equation driving the free electron density over time is [13]:

$$\partial_t \rho = W(I)(\rho_{at} - \rho) + \frac{\sigma_B(\omega)}{U_{IP}} \rho I - \alpha \rho^2 \tag{2.5}$$

where $W(I)$ is the photoionization rate, $I = \frac{1}{2}\epsilon_0 c E^2$ is the pulse intensity (if E is the electric field), $\sigma_B(\omega)$ is the inverse Bremsstrahlung cross section, $U_{IP} = \text{IP}$, ρ_{at} is the initial atomic density and α is the recombination coefficient. The ionization fraction of a gas is therefore given by:

$$\eta(r, t) = \frac{\rho(r, t)}{\rho_{at}} \tag{2.6}$$

The photoionization rate depends on the ionization regime. Indeed, this process can be linear if the energy of the photon E_{photon} is higher than the ionization potential (IP) of the system (typically $E_{photon} > 10$ eV or $\lambda < 124$ nm or $\omega > 2418$ THz, where $E_{photon} = hc/\lambda$ and $\omega = 2\pi c/\lambda$, c is the speed of light and h is the Planck constant) or nonlinear if the photon energy is lower than the ionization potential of the system ($E_{photon} < IP$). If the process is nonlinear, typically at infrared wavelength, it must be considered from three different regimes depending on the Keldysh parameter [14, 15]:

$$\gamma_K = \omega_0 \frac{\sqrt{2m_e U_{IP}}}{q_e E_{max}} \tag{2.7}$$

with E_{max} the electric field maximum amplitude, ω_0 the electric field frequency and m_e the electron mass, as introduced below.

2.2.1 *Multiphoton Ionization (MPI)*

This regime is dominant for “small” electric fields, i.e. large Keldysh parameter $\gamma_K \gg 1$. It is characterized by a power law dependence of the ionization rate on the laser intensity:

$$W_{MPI}(I) \propto \sigma_n I^n \tag{2.8}$$

where $n = mod\left\{\frac{U_{IP}}{\hbar\omega_0}\right\} + 1$ is the minimal number of photons required to ionize the system and σ_n is the n-photon absorption cross section. To get a feeling on the number of photons required for MPI in the nonlinear laser-gas interactions occurring in the context of this thesis, see Table 2.1.

Table 2.1 Number of photons required for MPI in different gases. The 1st ionization potential is given for each atom

Field wavelength (nm)	Photon energy (eV)	Krypton (14 eV)	Argon (15.8 eV)	Neon (21.6 eV)	Helium (24.6 eV)
1800	0.69	21	23	32	36
2100	0.59	24	27	37	42

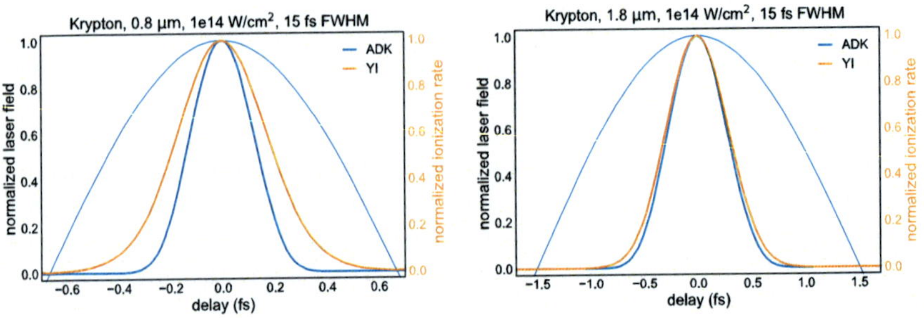

Fig. 2.3 Comparison of the instantaneous ionisation rates in krypton as a function of the laser field phase. Left: at 800 nm wavelength. Right: at 1800 nm wavelength. The pulse duration of 15 fs and the intensity of 10^{14} W/cm^2 are kept constant. Both quasi-static (ADK) and nonadiabatic (YI) model rates are shown, together with the laser field

2.2.2 *Tunneling Regime (TI)*

Tunneling ionization (TI) is the dominant regime when $\gamma_K < 0.5$ [16], but E_{max} is small enough such that the barrier-suppression regime (Sect. 2.2.4) is not yet reached. The strong electric field can bend the potential energy surface of the system and the electron is released through tunneling. Based on the Keldysh's theory [15], the so-called strong field approximation (SFA) considers the Coulomb potential as a perturbation and treats the laser field exactly. The 0th order SFA therefore totally neglects the potential and the ionization only takes place efficiently near the peak of the field (see Fig. 2.3 (blue curve)) [17]. This regime is characterized by an exponential dependence of the ionization rate on the instantaneous electric field [18]:

$$W_{\text{TI}}(E) \propto \exp(-\frac{2(2U_{\text{IP}})^{3/2}}{3E}) \tag{2.9}$$

2.2.3 *Tunneling in Molecules*

Because of the non-spherical molecular orbitals and the nuclear degrees of freedom, molecular ionization has additional features compared to atomic ionization. The complex molecular orbital symmetry properties, discussed in Sect. 2.1, usually lead the TI rate to be anisotropic. Moreover, it has been shown that the effect of lower-lying orbitals (HOMO-1, HOMO-2, …) cannot be excluded a priori [19], such that multi-channel strong field ionization [20] must be considered. From the intuitive picture of the outermost electron being loosely bound, molecules should have decreasing IP with increasing size. Yet, it has been demonstrated that a few organic molecules reach a full-ionization degree for intensities much higher than predicted by their TI rate

[21]. This apparent resistance to strong field ionization can be explained by different factors discussed in the thesis of Oppermann [11].

As an alternative ionization channel, TI from a lower-lying orbital (HOMO-1, HOMO-2, ...) should also be possible. When combining the Koopmans' theorem [22] with the single active electron approximation (SAE), in which the evolution of the remaining electrons is neglected [23], the distribution of the remaining electrons should not be modified. Therefore, excited states of the ion can be reached through ionization from lower-lying orbitals and shares the same orbital symmetry as its associated excited state in the neutral. The probability of reaching an excited state via this channel scales like the molecular orbital TI rate for the associated neutral orbital [11]. From another point of view and considering non-adiabatic electron dynamics during TI (SAE no more valid), the excited state of the ion can be associated with the electron hole moving to a lower-lying orbital during TI [24]. This mechanism is referred to as inelastic tunneling.

2.2.4 Barrier-Suppression Regime (OTBI)

This regime, also called over-the-barrier ionization (OTBI), is reached when the external field is strong enough such that there is no energy barrier separating the electron from the free space. It is characterized by a nearly linear dependence of the ionization rate on the incident electric field. The corresponding critical intensity can be approximated by [25]:

$$I_{\mathrm{OTBI}} \cong \frac{c{\epsilon_0}^3\pi^2{U_{\mathrm{IP}}}^4}{2Z^2{q_{\mathrm{e}}}^6} \tag{2.10}$$

where Z is the charge state of the corresponding atom or ion.

2.2.5 Ionization Rate Models

When $\gamma_{\mathrm{K}} \ll 1$, the last two regimes (TI and OTBI) are well described by the quasi-static approximation. The variation of the laser field is so slow that the instantaneous ionization rate coincides with a static one. This is in contrast with the multiphoton regime, where the rate exponentially decreases with the laser frequency. If the static ionization rate $W(E)$ is well described in the tunneling regime by the Ammosov-Delone-Krainov (ADK) formula [18] and has been experimentally verified for small molecules until $\gamma_{\mathrm{K}} \cong 3$ [26], this is not sufficient when looking at the intermediate regime as it is the case in the context of this thesis. For this reason, M. Yudin and M. Ivanov have developed a non-adiabatic model (YI, based on the Perelomov,

Popov and Terent'ev (PPT) tunneling model [27]) that allows calculating sub-cycle, time-dependent ionization rates for larger values of γ_K.

A comparison between the ADK quasi-static model and the YI non-adiabatic one is shown in Fig. 2.3 for two laser frequencies ($\lambda = 800$ nm and $\lambda = 1800$ nm). If both models follow a similar rate description in the low laser frequency ($\lambda = 1800$ nm) case where $\gamma_K = 0.48$, the divergence is obvious as the field frequency is increased and $\lambda = 800$ nm ($\gamma_K = 1.08$).

2.3 Propagation of Short and Intense Laser Pulses

While a typical low intensity laser beam propagates and diverges following the Gaussian optical theory, ultrashort ($\sim 10^{-15}$ s) IR pulses can maintain high intensity ($6 \cdot 10^{13}$ W/cm^2 in air) over multiple Rayleigh range z_R. Such phenomenon is called laser filamentation and implies a dynamical interplay between at least two nonlinear processes. Typically, and in the purpose of this thesis, the main effects are Kerr self-focusing, Gaussian diffraction, geometrical divergence and plasma defocusing.

Over the filament propagation, different effects occur that lead to a strong spatio-temporal reshaping of the pulse. Hence, the spectrum typically broadens and is modulated because of the interferances between the newly created frequencies. The temporal shape is also strongly modified as the refractive index changes over the intensity profile.

2.3.1 Kerr Self-focusing

The Kerr self-focusing effect is at the origin of the filament. It is the result of the transverse intensity profile of the beam and the nonlinear dependance on the medium refractive index:

$$n(r,t) = n_o + \Delta n_{\text{Kerr}}(r,t) \tag{2.11}$$

where Δn_{Kerr} directly rely on the Kerr index n_2 and the intensity I in the pulse by:

$$\Delta n_{\text{Kerr}}(r,t) = n_2 I(r,t) \tag{2.12}$$

The spatial intensity gradient of the pulse profile therefore induces an inhomogeneity of the refractive index that acts as a convex lens in the case of a Gaussian profile. In order to effectively focus the beam, such effect must be stronger than the competing defocusing one. In a loose focusing geometry, and for pulses in the short-wavelength infrared (SWIR) range that require high nonlinear processes to efficiently generate a plasma, the Gaussian diffraction is the first defocusing effect that appear. Therefore, the threshold value that a Gaussian laser pulse has to reach for the Kerr effect to equal the diffraction is the critical power:

$$P_{cr} \cong \frac{1.896\lambda^2}{4\pi n_0 n_2} \tag{2.13}$$

For a given wavelength λ and a pulse peak power, it is therefore possible to control the filamentation regime by selecting an adequate gas and adjusting the pressure p, assuming $n_2 \sim p/p_{atm}$.

2.3.2 *Plasma Defocusing*

From Eq. 2.5 driving the free electron density over time for a given ionization rate $W(I)$ (see Sect. 2.2), pulse intensity I, and recombination coefficient α, the plasma quantity $\rho(r, t)$ can be expressed in e$^-$/cm^3. Analogously to the Kerr nonlinear index $\Delta n_{Kerr}(r, t)$, the plasma defocusing index is given by [28]:

$$\Delta n_p(r, t) = -\frac{2\pi {q_e}^2 \rho(r, t)}{n_0 m_e \omega_0^2} \tag{2.14}$$

As the plasma is cumulative, the maximal free electron density is found at the trailing edge of the pulse and the defocusing effect is stronger for the delayed spectral components.

In addition to the plasma effect, the geometrical constraint plays an important role as shown in Fig. 2.4. By focusing a pulse with a 1.5 m focal lens, the filament regime holds over ~ 60 cm, while only ~ 5 cm distance is calculated for a 0.5 m focal lens.

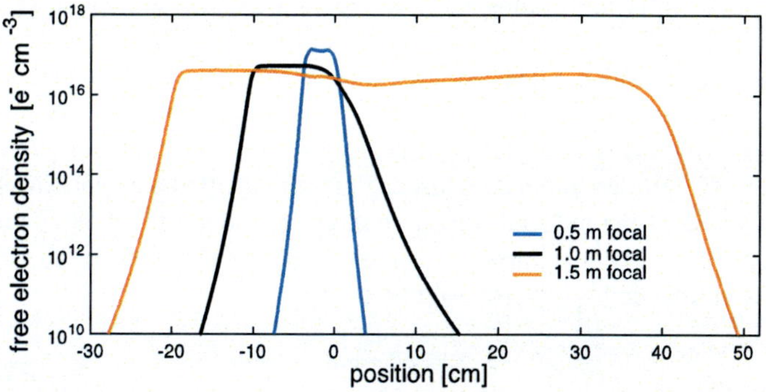

Fig. 2.4 Free electron density in the center of a same pulse (1.5 mJ, 60 fs, 3 mm diameter) focused with 0.5 m, 1 m and 1.5 m into 1 bar of Argon. The 0 position corresponds to the focus of each lens. Taken from [13]

From Fig. 2.4, the loose focusing geometry also allows to reduce the plasma density in the core of the filament.

2.3.3 Spectral Broadening

During filamentation, the spectrum is strongly broadened and can, in some cases, leads to a supercontinuum. In the framework of Fourier optics, spectral pulse broadening is attractive as it allows a band-limited pulse to be further compressed in time. The Kerr nonlinearity is directly responsible for this effect through self-phase modulation (SPM) [29, 30]. Because of the intensity dependence of the refractive index over time $n(t) = n_0 + n_2 I(t)$ (from Eq. 2.11), the variation of $n(I)$ is:

$$\frac{dn(I)}{dt} = n_2 \frac{dI}{dt} \tag{2.15}$$

which results in the phase shift:

$$\phi(t) = \omega_0 t + \frac{2\pi L}{\lambda_0} n(I) \tag{2.16}$$

where L is the propagation length. This phase shift, in turn, induces the pulse frequency to follow:

$$\omega(t) = -\frac{d\phi(t)}{dt} = \omega_0 - \frac{2\pi L n_2}{\lambda_0} \frac{dI}{dt} = \omega_0 - \Delta\omega \tag{2.17}$$

Thus, the leading edge of the pulse is redshifted (i.e. shifted toward the low frequencies), while the trailing edge is blue shifted (i.e. shifted toward the high frequencies). If the Kerr effect acts alone, the spectrum broadens symmetrically. Although, the refractive index also modifies because of higher order effects such as plasma defocusing that acts as a concave lens and is stronger at the trailing edge in reason of the density of plasma accumulated over the pulse duration. The blue shifted wavelengths are more defocused than the redshifted ones and thus propagates faster, because of the reduced refractive index $n(t)$ outer the core of the filament. This propagation effect is known as self-steepening and can drive the pulse to self-compress in time under specific conditions.

2.4 High Harmonic Generation (HHG)

When the regime of the interaction between the laser pulse and the atom is strongly nonperturbative ($\gamma_K \ll 1$), the dramatically nonlinear process of high-order harmonic generation (HHG) can occur. This process describes the conversion of an intense ultrashort laser pulse to integer multiples of its fundamental frequency.

The first observations of this process were made in Chicago (KrF laser's 17th harmonic) [31] and in Saclay (Nd:YAG laser's 33rd harmonic) [32] in the late 1980s. Following the perturbation theory, a typical HHG spectrum generated in a noble gas shows a rapid decrease of the low-order harmonics (below the ionization potential of the gas). When entering the non-perturbative regime, the nearly constant efficiency conversion forms a broad "plateau" that extends until the abrupt cut-off of the signal on the high energy side.

2.4.1 The Three-Step Model

The first theoretical framework for the description of the broad harmonic plateau was introduced in the early 1990s by Krause [33] and Corkum [34]. Based on a semiclassical theory, they proposed a "three-step model" in which an electron is first tunnel ionized by a strong laser pulse electric field and driven away from the nucleus. In a second time, when the oscillating electric field changes its sign, the electron is accelerated back towards the nucleus. Finally, the electron recombines with the atom and the significant amount of kinetic energy acquired from the electric field in the continuum cumulates to the ionization potential of the atom and is emitted by a high energy photon. Since the emission process occurs twice per optical cycle, it results in a comb of odd-order harmonics of the fundamental driving laser field in the frequency domain.

From this simple picture, the universal law for the energy cut-off value can be written [33]:

$$E_{\max} = U_{\mathrm{IP}} + 3.17 \cdot U_{\mathrm{p}} \tag{2.18}$$

where $U_{\mathrm{p}} \cong 9.33 \cdot \lambda^2 \cdot I_{\mathrm{S}}$ is the ponderomotive energy, i.e. the mean quiver energy of the electron in the laser field, U_{IP} is the ionization potential of the generation gas and I_{s} is the maximal intensity of the driving laser pulse during the interaction.

2.4.2 Wavelength Dependence and Strong-Field Approximation Model (SFA)

As the main interest of HHG in this thesis is to produce bright tabletop soft X-ray pulses for probing ultrafast molecular dynamics in the water window (284–543 eV), the cut-off energy and its dependencies are of most importance. If the single-atom harmonic cut-off energy is proportional to λ^2 and to the laser intensity I_s, a first approach to extend the energy is to increase the laser intensity. Unfortunately, due to the depletion of the ground state there is a limit beyond which the ionization is saturated and the laser intensity I does not contribute to further accelerate the electrons ($I = I_s$). Moreover, the plasma defocusing effect does not allow the intensity to be indefinitely increased in the core of the gas-field interaction. The alternative way to extend the cut-off energy is via the λ^2 scaling and then using longer-wavelength lasers. Nevertheless, the challenge of long-wavelength HHG driving is the drastically reduced single-atom yield s_q which scales as $\lambda^{-5.5}$ [35, 36]. Fortunately, this could be partially compensated by increasing the gas IP (i.e. going for another noble gas) which reduces the ground state depletion and allows for higher laser intensities with reduced plasma formation.

Analytically, the semiclassical approach of the three-step model was confirmed by a quantum treatment from M. Lewenstein et al. based on the strong-field approximation (SFA) and including the wave packet spreading and interference effects [17]. The depletion of the ground state was initially neglected, and added in a second publication [37]. The model relies on the SAE approximation and treats the free electron without the influence of the Coulomb potential.

The implementation of the model in a Python package allowed to efficiently simulate different conditions of generations and the results that accurately reproduced the experimental observations made in the context of this thesis are shown in Fig. 2.5. For a similar conversion efficiency, the model correctly predicts the extension of the cut-off when switching from neon (IP = 21.6 eV) to helium (IP = 24.6 eV) as generation gas. Additionally, when using a short laser pulse (Fig. 2.5 (right)), the model anticipates the strong dependence on the fundamental carrier envelope phase (CEP) value close to the cut-off region. As the effect of the Coulomb potential is not included, the model fails to correctly describe the harmonics at the low energy side of the spectrum.

The semiclassical model also predicts the dispersion of the recombination times as shown in Fig. 2.5 (bottom, right). This effect is called the attochirp and corresponds to a spectral group delay dispersion (GDD) of the emitted harmonics [38]. It is the main limitation to the duration of Fourier-synthesized pulses generated by HHG. As this chirp is almost constant for a given class of trajectories of the electrons in the continuum, it can in theory be compensated by propagating the pulses in a suitable dispersive gas [38].

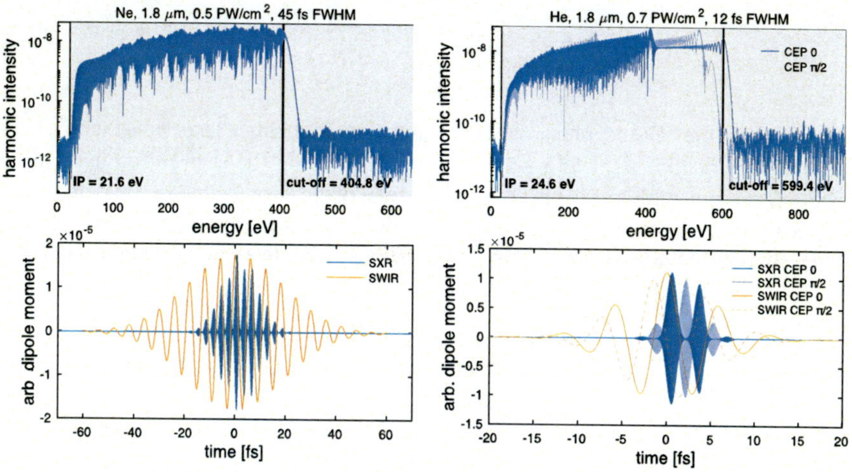

Fig. 2.5 Simulations of a single atom HHG process at 1800 nm for three generation conditions using the SFA model implemented with the YI ionization rate. Left: generation in neon with a 5×10^{14} W/cm^2 pulse intensity of 45 fs FWHM duration. Right: generation in helium with a 7×10^{14} W/cm^2 pulse intensity of 12 fs FWHM duration and two different carrier-envelope phase (CEP) for comparison. The top panels show the simulated HHG spectra with respective cut-off energies whereas the bottom panels show the temporal structures of the pulses. The gray shaded areas of the top panels explicit the validity range of SFA

2.4.3 Propagation Effects

In order to result in a detectable soft X-ray radiation, the HHG signal of the isolated emitters must coherently build up. Indeed, a satisfactory flux will arise only if good propagation conditions are achieved over the interaction's length. This macroscopic response [39–41] plays an important role in the description of the HHG in the far field, where they are used for spectroscopic measurements in this study. In addition to the specific effects of the HHG process discussed in this section, all the propagation effects of short and intense laser pulses discussed in Sect. 2.3 must be considered.

Of most importance, the phase matching provides the condition for which the phase front of the generated harmonic field is synchronized with the phase front of the fundamental laser field over the propagation axis z. Following Popmintchev et al. [42], the phase mismatch relation for the qth harmonic can be written as:

$$\begin{aligned}\Delta k_{\mathrm{q}} &= \left(k_{\mathrm{q}} - q k_0\right) - K_{\mathrm{q,dip}} \\ &= \Delta k_{\mathrm{q,geo}} + \Delta k_{\mathrm{q,el}} + \Delta k_{\mathrm{q,at}} - K_{\mathrm{q,dip}}\end{aligned} \tag{2.19}$$

with k_{q} and k_0 the wave vectors of the harmonic and of the fundamental laser fields. The coherence length $L_{\mathrm{q}}^{\mathrm{coh}} = \pi/\Delta k_{\mathrm{q}}$ being proportional to the inverse of the phase mismatch, efficient HHG is achieved if Δk_{q} is minimized. The geometric dispersion:

$$\Delta k_{\mathrm{q,geo}} = -q\frac{\partial}{\partial z}\left[\tan^{-1}\frac{z}{z_{\mathrm{R}}} - \frac{kx^2}{2R(z)}\right]^{x,z\to 0} \cong -\frac{q}{z_{\mathrm{R}}} \tag{2.20}$$

accounts for the "Gouy" phase shift introduced by focusing a laser pulse with a beam radius of curvature $R(z)$ to a small region in space. In the convention used in this section, the sign of this effect is always negative. In the strong-field regime, the single-atom induced dipole phase strongly depends on the laser intensity. Therefore, the Gaussian spatial intensity distribution results in longitudinal and transverse gradients of the dipole phase. Its contribution to the phase mismatch is:

$$K_{\mathrm{q,dip}} = \nabla\varphi_{\mathrm{q,dip}} \tag{2.21}$$

where the intrinsic dipole phase $\varphi_{\mathrm{q,dip}}$ is the action acquired by the electron during its trajectory in the laser field. In the harmonic spectral plateau region, this effect depends on the class of trajectories, "short" or "long", where the phase-matching conditions select one of these two trajectories.

As already introduced in Sect. 2.3.2, the free electron density $\rho(r,t)$ generated by the ionization process causes a modification in the refractive index. The associated phase mismatch is:

$$\begin{aligned}\Delta k_{\mathrm{q,el}} &= k_{\mathrm{q,el}}(r,z) - qk_{0,\mathrm{el}}(r,z)\\ &\approx -q\lambda_0 r_0\rho(r,t)\end{aligned} \tag{2.22}$$

where r_0 is the classical radius of the electron and the plasma dispersion for the harmonic field is neglected as the frequencies of high harmonics are much higher than the plasma frequency. Finally, $\Delta k_{\mathrm{q,at}}$ accounts for the neutral (atomic) dispersion mismatch. Considering Eq. 2.6 and $\rho_{\mathrm{at}} = p/k_{\mathrm{B}}T$ where k_{B} is the Boltzmann constant, T the gas temperature, and p the gas pressure, and assuming a non-guiding generation geometry (gas cell) with $K_{\mathrm{q,dip}} \approx 0$, these contributions lead to:

$$\Delta k_{\mathrm{q}} \approx \frac{qp}{k_{\mathrm{B}}T}\left(\frac{2\pi\Delta n_{0,\mathrm{q}}}{\lambda_0}(1-\eta(r,t)) - \lambda_0 r_0\eta(r,t)\right) - \frac{q}{z_{\mathrm{R}}} \tag{2.23}$$

Here, $\Delta n_{0,q}$ is the difference between the atomic refractive indices of the gas at the fundamental and qth harmonic wavelengths. Therefore, if the spatiotemporal reshaping effects discussed in Sect. 2.3 are ignored, the gas pressure p, the ionization fraction $\eta(r,t)$, and the Rayleigh range z_{R} are the only adjustable parameters to achieve good phase matching. Although, because of the time dependence of the ionization fraction $\eta(r,t)$ along the propagation, perfect phase matching can be achieved only over a finite propagation length [41]. Moreover, as a consequence of the phase-matching condition $n_{\mathrm{pm},0} \to 1$, the group velocity $v_{\mathrm{g}} = c(n - \lambda dn/d\lambda)^{-1}$ must also be considered, especially for short SWIR pulses [43]. Indeed, in contrast to the group velocity of the harmonics $v_{\mathrm{g,q}} = c$, the group velocity of the fundamental

Table 2.2 L^{abs} for typical HHG conditions over the water window at 292.3 K

Gas	Pressure (mbar)	L^{abs}_{280eV}	L^{abs}_{540eV}
Neon	750	1 mm	6 mm
Helium	4000	5 mm	>30 mm

$v_{g,0} = c$ cannot be achieved in a noble gas, because all the $\lambda_0 dn/d\lambda$ terms have the same sign [42]. Therefore, consistently to the coherence length L^{coh}, the walk-off length $L^{walk-off}$ is the distance between two emitters whose harmonic emission interferes destructively because of the group velocity mismatch [43]. This effect strongly depends on the pulse duration and can partially be compensated by the other propagation effects.

The last important effect of propagation is the gas absorption length L^{abs}, corresponding to the distance over which the intensity of the previously generated harmonics falls to $1/e$.

$$L_q^{abs} = \frac{1}{\rho_{at}\sigma_q} \tag{2.24}$$

where σ_q is the ionization cross section for the qth harmonic. With $\rho_{at} = p/k_B T$, this distance can be estimated using the X-ray absorption table from Henke et al. [44]. Table 2.2 provides the values of L^{abs} for the generation conditions used during this thesis.

Taking all these effects into account, it is possible to build a macroscopic HHG response using a simple one-dimensional model. The generated signal in a static gas cell of length L is the coherent sum over all single atom emitters s_q (see Sect. 2.4.2) and can be written as [41]:

$$\begin{aligned} S_q &\propto \left| s_q \int_0^L \mathrm{d}z \exp\big(i\left(\Delta k_q + i\rho_{at}\sigma_q\right)(L-z)\big)\right|^2 \\ &= \left|s_q\right|^2 e^{-\rho_{at}\sigma_q L} \cdot \frac{\cosh \rho_{at}\sigma_q L - \cos \Delta k_q L}{\Delta k_q^2 + (\rho_{at}\sigma_q)^2} \end{aligned} \tag{2.25}$$

2.5 Soft X-Ray Absorption Spectroscopy

As introduced in Sect. 2.1, every molecular system contains many electronic transitions. While outer lying molecular orbital transitions typically corresponds to the energy of photons in the ultraviolet to infrared wavelength range, promotion or ionization of an electron from a core shell require much higher energies. As the HHG conversion process introduced in the previous section allows reaching these core energies for some atomic species, using HHG to probe the molecular structure from core shell electrons is accessible for energies up to 540 eV. A list of all electronic transitions spanning from 150 to 540 eV is given in the Appendix A.1. Also, in reason for the sub-femtosecond duration of the generated SXR pulses and the intrinsic temporal synchronization with the driving laser, combining both fundamental and HHG pulses permits the measurement of ultrafast ($<10^{-14}$ s) dynamics. During this thesis, the dynamics are probed using the X-ray absorption spectroscopy (XAS) technique in the transmission geometry.

2.5.1 The Absorption Cross-Section

In order to understand and characterize radiative processes, the transition probability $M_{\mathrm{i}\to\mathrm{f}}$ from an initial state Ψ_{i} to a final state Ψ_{f} introduced in Sect. 2.1.4 and given by Eq. 2.4, must be related to experimental observables. The oscillator strength is given by [45]:

$$f_{\mathrm{if}} = \frac{2m_{\mathrm{e}}}{3\hbar}\Delta E_{\mathrm{if}} M_{\mathrm{if}} \tag{2.26}$$

where ΔE_{if} is the energy required to promote an electron from the initial state to the final state. In the case of a photoabsorption process, it corresponds to the energy of the absorbed photon $\Delta E_{\mathrm{if}} = h\nu$.

On the other side, the experimental measurement of a sample absorbance $A = \ln(I_{\mathrm{ref}}/I)$ (see Sect. 3.5.6) relates to the absorption cross-section σ_{if} [cm^2/molec] with:

$$A = \sigma_{\mathrm{if}} L \rho_{\mathrm{mol}} \tag{2.27}$$

where L [cm] is the absorption path length given by the sample thickness, and ρ_{mol} [molec/cm^3] is the molecular density. Finally, the absorption cross-section is linked to the oscillator strength by:

$$\begin{aligned} \int \sigma_{\mathrm{if}}(E)\mathrm{d}E &= \frac{\pi q_{\mathrm{e}}^2}{m_{\mathrm{e}}c} f_{\mathrm{if}} \\ \sigma_{\mathrm{if}}(E) &= \frac{\pi q_{\mathrm{e}}^2}{m_{\mathrm{e}}c} L(E - \Delta E_{\mathrm{if}}) f_{\mathrm{if}} \quad \text{with} \quad \int L(E - \Delta E_{\mathrm{if}})\mathrm{d}E = 1 \end{aligned} \tag{2.28}$$

where the Lorentzian function $L(E - \Delta E_{\mathrm{if}})$ provides an accurate approximation to the actual line profile. The intrinsic width of the absorption line reflecting the uncertainty in the transition energy due to the finite lifetimes of the excited state.

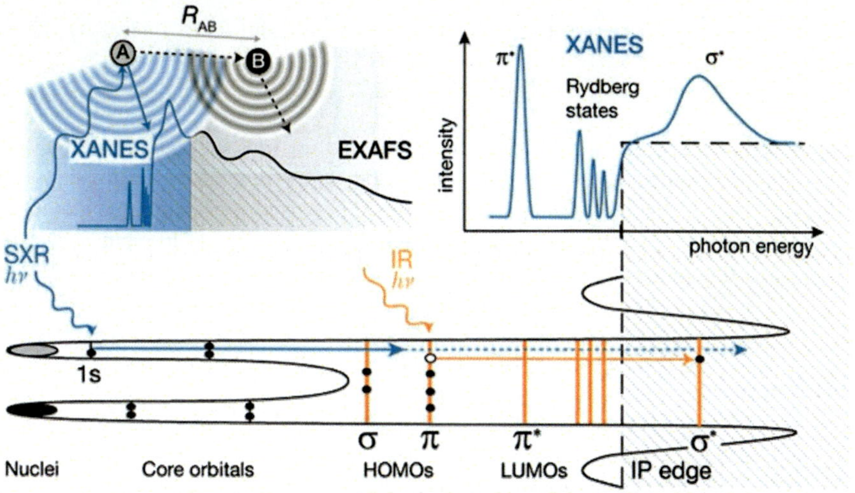

Fig. 2.6 Schematic representation of a XAS measurement. Top left: the complete measured spectrum is divided into two regions. The XANES provides a direct measure of the absorption cross-section, whereas the EXAFS shows oscillations induced by the scattering of the ionized photoelectron from neighboring atoms. Top right: close-up on the XANES region related to the schematic potential of a diatomic molecule (bottom). The main molecular orbitals features are represented in both the schematic potential (bottom) and the close-up (top right). The blue color refers to SXR absorption induced transitions and the orange color refers to the IR pulse possible excitations

2.5.2 X-Ray Absorption Near-Edge Structure (XANES)

The X-ray absorption near-edge structure (XANES) introduced in the early 1980s and also called near-edge X-ray absorption fine structure (NEXAFS), is the spectrum region close to the ionization threshold energy edge. The structure of the spectrum is dominated by the core transitions to unoccupied molecular orbitals and to broad shape resonances lying above the ionization potential as shown in Fig. 2.6. It provides a direct measurement of the absorption cross-section and allows to energetically resolve the element-specific localized molecular structure.

2.5.3 Extended X-Ray Absorption Fine Structure (EXAFS)

In the high kinetic energy range of the photoelectron, the absorption spectrum is dominated by the extended X-ray absorption fine structure (EXAFS) where the scattering of the photoelectron from neighboring atoms results in oscillatory modulations [46]. Its analysis can yield the number of atoms in the immediate environment of the absorbing one and provides accurate internuclear distances.

The complete physical description when looking at standardized data $\chi(E)$ is given in the k-space by the EXAFS equation [47]:

$$\chi(k_e) = \sum_j \frac{N_j f_j(k_e)}{k_e R_j^2} \exp(-2k_e{}^2 \sigma_j^2) \exp\left(-\frac{2R_j}{\lambda_j(k_e)}\right) \sin(2k_e R_j + \delta_j(k_e)) \quad (2.29)$$

where $k_e = \sqrt{2m_e(E - IP)/\hbar^2}$ is the photoelectron wavenumber, N_j the atomic coordination number, R_j the interatomic distance, σ_j^2 represents the mean-square disorder in the distance for the j th shell, $\lambda_j(k_e)$ is the photoelectron inelastic mean free path, f_j is the photoelectron back-scattering amplitude and $\delta_j(k_e)$ the corresponding phase.

References

1. Daintith J (1966) Dictionary of chemistry, 6th edn. Elsevier
2. Winter MJ (2016) Chemical bonding. Oxford University Press
3. Atkins PW, Friedman R (2011) Molecular quantum mechanics, 5th edn
4. Miessler GL, Tarr DA, Kauffman GB (2000) Inorganic chemistry, 2nd edn, vol 77, no 2
5. Atkins BPW, Child MS, Phillips CSG (2006) Tables for group theory
6. Ying JF, Leung KT (1994) Absolute transition probability measurement of valence-shell electronic structure of CF 4 and CCl 4: Bethe surfaces, and dipole-dominated preionization-edge valence and Rydberg states. J Chem Phys 100(10):7120–7134
7. List NH, Kauczor J, Saue T, Jensen HJA, Norman P (2015) Beyond the electric-dipole approximation: a formulation and implementation of molecular response theory for the description of absorption of electromagnetic field radiation. J Chem Phys 142(24):244111
8. Bandrauk AD (1995) Molecules in laser fields. In: Frontiers of chemical dynamics. Springer Netherlands, Dordrecht, pp 131–150
9. Franck J, Dymond EG (1926) Elementary processes of photochemical reactions. Trans Faraday Soc 21:536
10. Condon EU (1928) Nuclear motions associated with electron transitions in diatomic molecules. Phys Rev
11. Oppermann M (2011) Resolving strong field dynamics in cation states of CO2 via optimised molecular alignment. Imperial College London
12. Codling K, Frasinski LJ (1994) Coulomb explosion of simple molecules in intense laser fields. Contemp Phys 35(4):243–255
13. Berti N (2016) Champ fort et rétro-propagation en filamentation Laser. University of Geneva
14. Faisal FHM (1973) Multiple absorption of laser photons by atoms. J Phys B At Mol Phys 6(4):L89–L92
15. Reiss HR (1980) Effect of an intense electromagnetic field on a weakly bound system. Phys Rev A 22(5):1786–1813
16. Ilkov FA, Decker JE, Chin SL (1992) Ionization of atoms in the tunnelling regime with experimental evidence using Hg atoms. J Phys B At Mol Opt Phys 25(19):4005–4020
17. Lewenstein M, Balcou P, Ivanov MY, L'Huillier A, Corkum PB (1994) Theory of high-harmonic generation by low-frequency laser fields. Phys Rev A 49(3):2117–2132
18. Ammosov MV, Delone NB, Krainov VP (1986) Tunnel ionization of complex atoms and atomic ions in electromagnetic field. In: Soviet physics—JETP, p 138
19. Jia X, Li W, Liu J, Chen J (2009) Alignment-dependent nonsequential double ionization of N2 in intense laser fields: the role of different valence orbitals. Phys Rev A 80(5):053405
20. Smirnova O et al (2009) High harmonic interferometry of multi-electron dynamics in molecules. Nature 460(7258):972–977
21. Hankin SM, Villeneuve DM, Corkum PB, Rayner DM (2001) Intense-field laser ionization rates in atoms and molecules. Phys Rev A 64(1):013405

22. Koopmans T (1934) Über die Zuordnung von Wellenfunktionen und Eigenwerten zu den Einzelnen Elektronen Eines Atoms. Physica 1(1–6):104–113
23. Krausz F, Ivanov M (2009) Attosecond physics. Rev Mod Phys 81(1):163–234
24. Torlina L, Ivanov M, Walters ZB, Smirnova O (2012) Time-dependent analytical R-matrix approach for strong-field dynamics. II. Many-electron systems. Phys Rev A 86(4):043409
25. Brabec T, Krausz F (2000) Intense few-cycle laser fields: Frontiers of nonlinear optics. Rev Mod Phys 72(2):545–591
26. Cornaggia C, Hering P (2000) Nonsequential double ionization of small molecules induced by a femtosecond laser field. Phys Rev A 62(2):023403
27. Perelomov A, Popov V, Terent'ev M (1966) Ionization of atoms in an alternating electric field: II. Sov J Exp Theor Phys
28. Kosareva O et al (2011) Arrest of self-focusing collapse in femtosecond air filaments: higher order Kerr or plasma defocusing? Opt Lett 36(7):1035
29. Shen YR (1985) The principles of nonlinear optics, no 2
30. Agrawal G (2001) Nonlinear fiber optics. Elsevier Science
31. McPherson A et al (1987) Studies of multiphoton production of vacuum-ultraviolet radiation in the rare gases. J Opt Soc Am B 4(4):595
32. Ferray M, L'Huillier A, Li XF, Lompre LA, Mainfray G, Manus C (1988) Multiple-harmonic conversion of 1064 nm radiation in rare gases. J Phys B At Mol Opt Phys 21(3):L31–L35
33. Krause JL (1992) High-order harmonic generation from atoms and ions in the high intensity regime. Phys Rev Lett 68(24):3535–3538
34. Corkum PB (1993) Plasma perspective on strong field multiphoton ionization. Phys Rev Lett 71(13):1994–1997
35. Tate J, Auguste T, Muller HG, Salières P, Agostini P, DiMauro LF (2007) Scaling of wave-packet dynamics in an intense midinfrared field. Phys Rev Lett 98(1):13901
36. Driever S et al. (2013) Tunable 1.6–2 μm near infrared few-cycle pulse generation by filamentation. Appl Phys Lett 102(19):191119
37. Antoine P, L'Huillier A, Lewenstein M, Salières P, Carré B (1996) Theory of high-order harmonic generation by an elliptically polarized laser field. Phys Rev A 53(3):1725–1745
38. Cousin SL et al (2017) Attosecond streaking in the water window: a new regime of attosecond pulse characterization. arXiv:1710.02208 [physics.a]
39. Balcou P, Salières P, L'Huillier A, Lewenstein M (1997) Generalized phase-matching conditions for high harmonics: the role of field-gradient forces. Phys Rev A 55(4):3204–3210
40. Dachraoui H et al (2009) Interplay between absorption, dispersion and refraction in high-order harmonic generation. J Phys B At Mol Opt Phys 42(17):175402
41. Heyl CM, Arnold CL, Couairon A, L'Huillier A (2017) Introduction to macroscopic power scaling principles for high-order harmonic generation. J Phys B At Mol Opt Phys 50(1):013001
42. Popmintchev T et al (2009) Phase matching of high harmonic generation in the soft and hard X-ray regions of the spectrum. Proc Natl Acad Sci 106(26):10516–10521
43. Hernández-García C et al (2016) Group velocity matching in high-order harmonic generation driven by mid-infrared lasers. New J Phys 18(7):073031
44. Henke BL, Gullikson EM, Davis JC (1993) X-ray interactions: photoabsorption, scattering, transmission, and reflection at E = 50–30,000 eV, Z = 1–92. At Data Nucl Data Tables 54(2):181–342
45. Okabe H (1979) Photochemistry of small molecules. Synth React Inorg Met Chem 9(4):393–394
46. Newville M (2001) EXAFS analysis using FEFF and FEFFIT. J Synchrotron Radiat 8(2):96–100
47. Teo BK (1986) EXAFS: basic principles and data analysis, vol 9. Springer, Berlin Heidelberg

Chapter 3
Method and Setup Description

This chapter introduces in detail the different elements assembled and designed to get the soft X-ray transient absorption experiment ready. Some maintenance routines and operational specifications naturally stem from the laser source's reliance on a commercial system. It describes and details every pulse involved in the experiment, as well as the optical post-treatments applied to the pump pulse and the high harmonic generation driver to increase the system overall performances. The condition of soft X-ray generation in helium with long wavelength lasers is discussed and compared to the most advanced similar systems, providing an overview of the state of the art for tabletop time-resolved absorption spectroscopy in the water window. The characteristics of the flat-field vacuum spectrometers that were used to measure the signals over the full water window are also presented and discussed in detail. Finally, the data acquisition and data analysis methods are provided.

3.1 Introduction

As required by the transient absorption scheme, a pump and a probe pulse are required to induce a perturbation on the sample and observe the following dynamics. These two pulses must share a known time relation, at least sometimes, to provide time resolution. In typical tabletop pump-probe experiment, the two pulses are extracted from the same oscillator source. It is the case of the setup presented here. Nevertheless, similar experiments rely on post synchronized pump-probe pulses as in the case of X-ray free electron laser (XFEL).

High-order harmonic generation (HHG) based sources have recently achieved unprecedented, sub-femtosecond, temporal resolution in transient absorption measurements [1–3]. While the pioneering HHG experiments were performed using near-infrared (NIR) laser-based sources with a cut-off limited to the sub-100 eV range, most spectroscopic studies providing both attosecond time resolution and

C. Schmidt, *Time-Resolved Soft X-Ray Absorption Spectroscopy of Molecules in the Gas and Liquid Phases*, Springer Theses,
https://doi.org/10.1007/978-3-030-67838-8_3

atomic spatial resolution require core electron excitation at significantly higher energies. In particular, the dynamics of organic and biological molecules would be optimally addressed in the so-called water transparency window, which lies between the carbon and oxygen K-shell absorption edges (commonly 284–543 eV) enabling X-ray absorption measurements of aqueous solutions [4].

The recent advancements in high-power laser/parametric sources have achieved sufficient photon flux (kHz repetition rates) for performing time-resolved transient absorption measurements in the water window range [5] and techniques, such as attosecond streaking [6–8], have recently been extended to this spectral range enabling the generation of the shortest attosecond pulses ever produced [9].

In the framework of this thesis, an SXR source was extended beyond the oxygen K-edge and can now reach 570 eV. The source is driven by 1.8 μm short-wavelength infrared (SWIR) pulses generated in an optical parametric amplifier (OPA) that are now post-compressed in two filamentation stages (Fig. 3.1). Long wavelength driving pulses are attractive for pulse post-compression in a filament, as the critical power P_{cr} scales with λ^2 [10, 11]. As the wavelength increases, the filament plasma density decreases and its role in counterbalancing Kerr self-focusing decreases accordingly. The long wavelength driving pulses used in this setup thus allow efficient and scalable filament-based pulse compression as discussed in Chap. 5 without the detrimental effects of plasma formation.

While broadband optical parametric chirped-pulse amplifiers (CPA) based on BiBO nonlinear crystals [12] or parametric amplification in the Fourier-plane [13] have been previously successfully demonstrated, the setup developed in this thesis

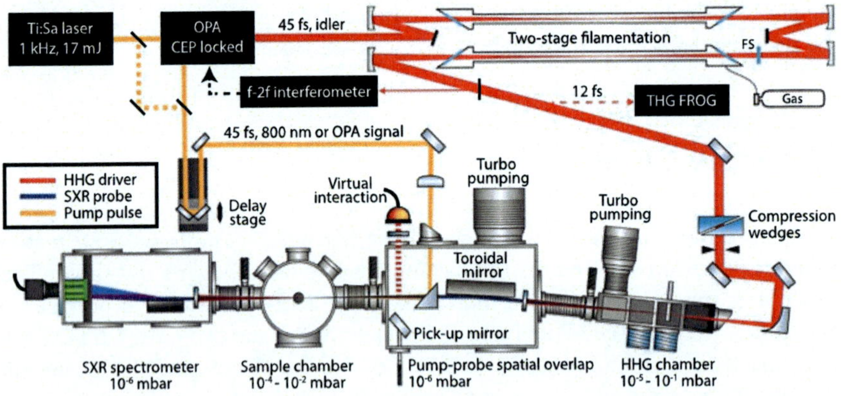

Fig. 3.1 General layout of the experimental setup. The 1.8 μm idler pulses (in red) are broadened by two successive filament lines of 2 m each and compressed in a fused silica plate before being focused into a gas target to drive the HHG process. The SXR beam is then refocused colinearly with the NIR pump pulse into the sample. After filtering the residual pump and driver radiation by multiple metallic filters, the transmitted SXR is measured by a custom spectrometer. Adapted from K. Zinchenko, ETH

shows that decent flux HHG can be achieved using a conventional BBO based Type-II optical parametric amplifier and a simpler and robust pulse compression method. Pulse compression together with the use of helium, a higher *IP* HHG gas target as compared to neon allow for SXR cut-off extension and improved temporal resolution in pump-probe experiments.

3.2 Laser Systems

The optical pulse driving the SXR generation and the one used to pump the sample are based on the same commercial femtosecond-laser system. A Ti:Sa oscillator (Vitara-T from Coherent) generates 480 mW of a 110 nm FWHM bandwidth pulse train. The central wavelength is 800 nm and the repetition rate is 80 MHz. Such bandwidth is capable of supporting <8 fs pulse duration.

The beam is then pre-amplified within a regenerative amplifier (Hidra-25 USP Hybrid HE). In a first stage, the pulse is stretched in time in order to avoid any high-intensity nor nonlinear effect during and after amplification. The chirped pulse train is then resampled to 1 kHz by an electro-optic Pockels cell and injected into a previously pumped Ti:Sa rod. By the mean of a resonator geometry, the pulse typically undergoes 15 round trips through the gain medium, being amplified to about 4 W, respectively 4 mJ. The pulse is then extracted from the resonator by a second electro-optic switch and subsequently amplified in a cryogenically cooled two-pass amplifier (Legend USP Cryo Amplifier from Coherent) pumped by two 45 W pulsed green lasers. After final amplification, the available energy is 23 mJ. Before being released from the laser box, the pulse is compressed to 45 fs FWHM and reaches maximum 17 mJ.

This intense beam can be split by an 80:20 beam splitter to get 800 nm photons available for pumping the sample. Most of the pulse energy is used to pump a BBO-based Type-II parametric amplifier (HE-TOPAS-Prime-Plus from Light Conversion) seeded with a white-light supercontinuum derived from the same pump pulse. The parametric amplifier output wavelength is tunable around the crystal degeneracy at 1600 nm from 1160 to 2480 nm. The energy conversion from parametric amplification is shown for the signal and the idler in Fig. 3.2.

The idler pulse is dedicated to the high-harmonic generation process while the signal is either used to pump the measured sample or directly dumped. The signal is separated from the idler by a high reflection polarized beam splitter followed by two high-reflection mirrors centered around the idler wavelength and having extremely low reflectivity over the signal wavelength range. These two mirrors can easily be replaced to fit the OPA set idler wavelength (see Sect. 3.3.1). The multistage amplification design coupled to the narrow phase-matching band given by type-II crystals seeded away from degeneracy implies to apply a temporal delay between signal and idler before the last amplification stage to avoid interferences. Such delay is easily realized as signal and idler propagate with orthogonal polarization in a case of type-II phase-matching. The last amplification stage pump delay is therefore

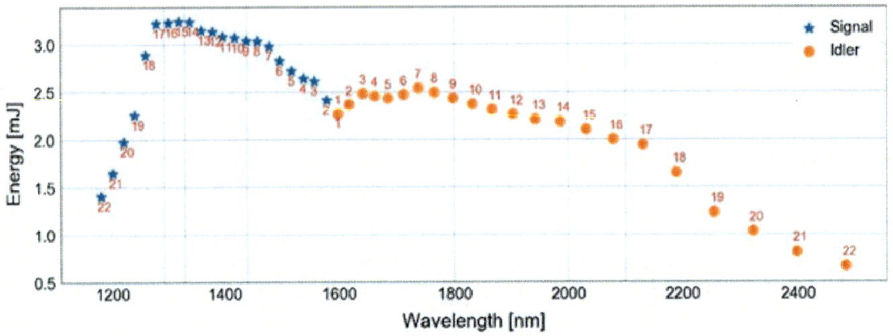

Fig. 3.2 Parametric amplifier output energies over the full range of signal and idler wavelength. The system is pumped at 1 kHz by 15 mJ of 800 nm, 45 fs pulses

optimized for a better idler pulse spatial mode and highest power. As a result, the corresponding signal beam is affected by temporal and spatial chirp [14]. At this stage, the idler is p-polarized and the signal is s-polarized.

Because of mechanical constrains and thermal fluctuations of the laboratory building, the laser specifications are subject to small drifts that must be measured and adjusted for good operation. Typically, the built-in laser chain requires a power check and small adjustments every day while a full realignment is necessary every week of measurement.

3.2.1 OPA Signal and Idler Pulse Characteristics

At the output of the Ti:Sa CPA laser, and for a fixed OPA setting, there are 3 available pulses. The 800 nm pump pulse (extracted before OPA), the SWIR idler pulse and the corresponding NIR signal pulse. Depending on the experiment requirement, it is therefore possible to choose the best combination. Figure 3.3 shows the available

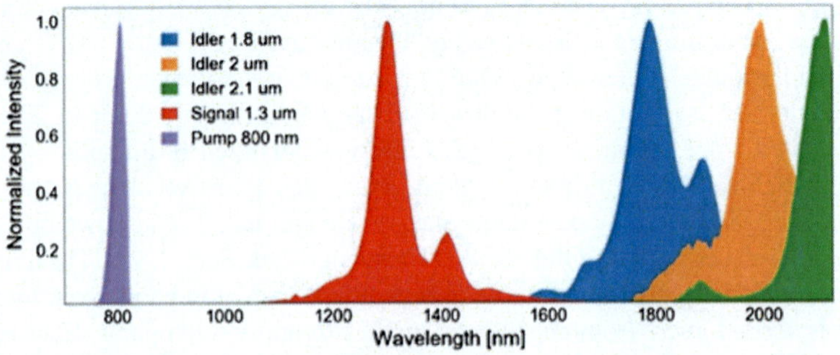

Fig. 3.3 OPA output pulse spectra measured with a NIRQuest spectrometer (by Ocean Optics). Only the 1.3 μm signal (2.1 μm idler) is presented as it is the only one used to pump the sample

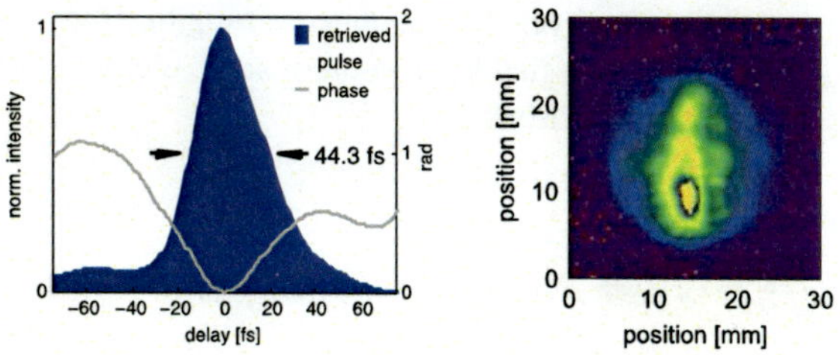

Fig. 3.4 Uncompressed 1.8 μm idler retrieved temporal intensity profile and phase (left) and spatial mode (right). The scale of the spatial profile may be not accurate

spectra used in this work to drive HHG or pump samples. The retrieved temporal profile as well as the temporal phase of the 1.8 μm idler at the OPA output is shown in Fig. 3.4 . Its typical spatial intensity profile is measured by 2 photons absorption on a CMOS camera after slightly reducing the beam size to fit the camera sensor.

3.2.2 Procedure for CPA-Compressor Grating Cleaning

On the side of the common transmissive and reflective optics cleaning methods described in every system manuals, the surface of the first grating in the Cryo two-pass amplifier compressor is subject to coating contamination at the center of the output beam position that tends to considerably reduce the output power and the spatial mode quality. Since the optical surface is made of epoxy and without overlying protection, any standard mechanical cleaning technique will fail and create permanent damages.

In a first step, multiple pinholes are set as references in order to mount the grating back precisely at the same position. After a very careful removal of the optic, the surface is first washed with clean water and a little dishwashing soap in a small container. By slowly bouncing the container, most of the grease and dirt are detached from the surface. Once most parts of the spot are removed, the grating is rinsed with distilled water. In a second time and only if the first cleaning failed to remove all the dirt, the surface is doused parallel to the groove orientation with small droplets of solvent using a bottle of compressed air crossed with the nozzle output which sprays ethanol. This process had to be done in total 3 times during the period of this work in order to maintain the 74% compressor transmission.

3.3 OPA Pulses Optimization

In a preliminary version of the setup, and as the neon generated SXR was able to cover the carbon K-edge, the 800 nm pump pulse and the 1.8 μm idler pulse were used as they are delivered by the commercial laser line. These pulses were later reshaped in space and time in order to reach higher HHG efficiency and cutoff and improve the time resolution of the transient absorption signal. As these post-treatments are part of the improvements obtained during this thesis, the results are discussed in the Chap. 5 and only the description of setup elements and the method are discussed in the following sections.

3.3.1 Idler Pulse Broadening and Compression

In the process of high-harmonic generation, a well-established technique to extend the generated energy cut-off is to use longer wavelength (λ) driving pulses [15, 16], taking advantage of the well known λ^2 dependence of the cut off frequency [17]. With the OPA idler set over 1.8 μm, it is possible to extend the SXR spectrum beyond the oxygen K-edge energy (535 eV), meaning a full coverage of the SXR water window [15]. The challenge of long-wavelength HHG driving is the drastically reduced single-atom yield which scales as $\lambda^{-5.5}$ [18]. However, this partially can be compensated using favorable phase-matching conditions and increased medium density. The use of short few-cycle duration pulses is highly desirable and allows HHG in the loose-focusing geometry thus increasing efficiency and compensating for the single-atom yield decrease.

During this thesis, the OPA idler pulses were post-compressed with a filamentation setup. As the critical power P_{cr} also scales with λ^2 [10], long wavelengths are adapted for non-destructive filamentation broadening. When the wavelength increases, the low-plasma density ($10^{15}-10^{17}$cm^{-3}) decreases and its role in counterbalancing Kerr self-focusing decreases accordingly [19].

The spectrum of the idler pulse is broadened by loosely focusing the beam using a 2 m focal length spherical mirror in a first 3.5 m long argon or krypton gas cell isolated by two 3-mm-thick CaF_2 windows cut at Brewster's angle for the first stage of spectral broadening via self-phase modulation. The beam is then collimated using a second 2 m focal length silver mirror and the residual chirp is compensated in a 1 mm thick fused silica glass plate (Fig. 3.5, Table 3.1). The pulse after the first compression stage is then launched into a second pulse compression stage consisting of an identical argon- or krypton-filled cell for subsequent spectral broadening. The

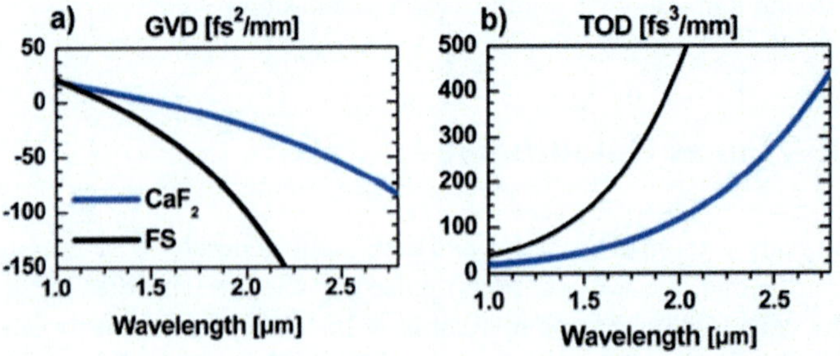

Fig. 3.5 GVD (**a**) and TOD (**b**) dispersion curve for CaF2 and Fused Silica. Reprinted from [21]

Table 3.1 GVD (fs2/mm) values of the most commonly used materials at the considered wavelength

	CaF_2	Fused silica
800 nm	27.873	36.163
1.3 μm	9.1167	−2.3731
1.8 μm	−10.870	−62.986
2 μm	−21.178	−100.61
2.1 μm	−27.016	−123.48

gas pressure in each cell can be independently controlled ranging from vacuum to 3 bar of absolute pressure. The two-stage pulse compression arrangement allows achieving a higher compression ratio and cleaner compressed pulses than a single stage, as originally demonstrated for a Ti:Sa laser pulses [20]. The beam is then collimated by a 3 m focal length silver mirror and the chirp is compensated again using a set of rotatable fused silica flat windows for fine adjustment of the dispersion.

At wavelengths longer than 1.3 μm, Fused Silica as many of the standard optical glasses become anomalously dispersive as shown in Fig. 3.4. The material group velocity dispersion GVD (fs^2/mm) of a medium with thickness L (mm) modifies the pulse duration. The induced pulse width Δt of a pulse with an original duration Δt_0 at FWHM, can be estimated by using:

$$\Delta t = \Delta t_0 \left[\left(1 + \frac{4ln2 \cdot GVD \cdot L \cdot C}{\Delta t_0{}^2} \right)^2 + \left(\frac{4ln2 \cdot GVD \cdot L}{\Delta t_0{}^2} \right)^2 \right]^{\frac{1}{2}} \tag{3.1}$$

where C is a chirp parameter defining the amount of initial positive chirp in the incoming pulse [22]. The value of the parameter C can be estimated with

$$C = \left[1 + \left(\frac{t_0}{t_{\min}} \right)^2 \right]^{\frac{1}{2}} = \left[1 + (t_0 \Delta\omega)^2 \right]^{\frac{1}{2}} \tag{3.2}$$

if $t_{\min}$ is the bandwidth-limited pulse duration.

During alignment, and pressure adjustment, the broadened spectrum is directly monitored using an InGaAs spectrometer while the fused silica flat window compression is optimized by directly looking at the generated SXR spectrum. The temporal characterization is performed on a separated beam line, picking the pulse just before focusing for HHG, using dispersion-free third-harmonic generation optical gating (THG-FROG).

While the full description of the compressed pulse is given in Sect. 5.5, the typical characteristics of the 1.8 μm pulse are his peak power of as much as 0.2 TW, for an average power of 2.1 W and a measured pulse duration down to 11.8 fs (sub-2 cycles). The losses in the whole compression process do not exceed 15%, mainly due to Fresnel losses in the cell windows and reflectivity of silver mirrors. Note that broadband reflective mirrors were used in order to assure wavelength flexibility of the setup. A remarkable improvement of the spatial beam profile is observed after the 2-stage filamentation thanks to the self-cleaning effect [23], which mainly originates from the spatial dependence of the Kerr induced non-linear refractive index of the incoming beam [24, 25].

3.3.2 Dispersion-Free Third-Harmonic FROG

The pulse temporal characterization is performed by a home-made frequency resolved optical gating (FROG) setup adapted to be fully reflective. It uses the third harmonic (THG) signal $E_{\text{sig}}^{\text{THG}}(t, \tau) = E^2(t)E(t-\tau)$ of a pellicle thin plate in order to measure very short and broadband SWIR pulses. The advantages of the THG-FROG signal are its overall sensitivity and the non-ambiguous phase sign allowing for a direct and complete chirp and third order information [26]. Moreover, typical spectrometers based on silicon CCD array detector cut at 1.1 μm and therefore don't cover the whole second harmonic signal. Nevertheless, swapping to a SHG configuration $E_{\text{sig}}^{\text{SHG}}(t, \tau) = E(t)E(t-\tau)$ is straight forward and was used in the present work to optimize the symmetry of the alignment, as the measured autocorrelation has to be symmetric in time. The scheme of the setup is shown in Fig. 3.6.

The fundamental principle of the autocorrelation measuring method is to record a signal generated by the interaction of a pulse with itself. The signal is typically generated by a known nonlinear interaction, ideally much faster than the pulse duration. The temporally gated pulse $E(t)$ spectrum is measured as a function of the delay τ between the two pulses giving a 2D signal, known as the spectrogram of the pulse:

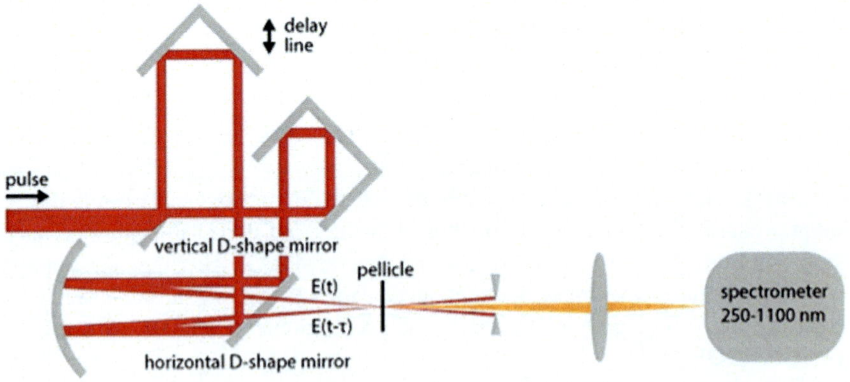

Fig. 3.6 Scheme of the THG-FROG used to characterize SWIR pulses. The initial transmissive beam splitters are replaced by two D-shape mirrors in order to avoid any dispersion of the short pulses during measurement. The THG medium is a 2 μm thick pellicle beam splitter that can be easily replaced by a BBO crystal for SHG-FROG configuration

$$I_{\text{FROG}}(\omega, \tau) = \left| \int E_{\text{sig}}(t, \tau) e^{-i\omega t} dt \right|^2 \tag{3.3}$$

This measurement, by the aim of a fast iterative retrieval algorithm, is sufficient to completely determine $E(t)$ [27].

3.3.3 f-2f CEP Stabilization

As the transient absorption spectroscopy signal is obtained by comparing the HHG spectrum through the sample with a reference one, the pulse stability over the time scale of the measurement is crucial in order to see the minor absorption features. In the few-cycle regime of the compressed idler pulse driving the SXR generation, the stability of the spectrum depends not only on the intensity of the source, but also on its carrier envelope phase (CEP) [28] which, if well controlled, can be used for the production of isolated attosecond pulses [9, 29].

To stabilize the phase of the idler pulses, a home-built active CEP control scheme was installed [30]. Relying on the phase relation between the white light supercontinuum and the idler pulses $\phi_{\text{idler}} = \phi_{\text{pump}} - \phi_{\text{wl}} - \frac{\pi}{2}$ due to the difference frequency generation in the OPA, the idler CEP is passively stable. The residual phase variation due to small drifts in the different optical path lengths of the seed and pump pulses can therefore be compensated by an active delay line on the high energy pump path. A piezo mirror-shifter STr-25 (from Piezomechanik) is mounted on the last flat mirror of the pump line and is activated by an SVR 150/1 piezo amplifier. The feedback command loop is controlled via a dedicated LabView interface at a maximum refresh frequency of 50 Hz. To perform the active stabilization parallel to the experiments, a weak reflection of the last filament cell Brewster window is sent to the optical setup shown in Fig. 3.7.

The idler low frequency CEP variations are measured by an f-to-2f interferometer where the second harmonic generated frequency $I_{SHG}(\omega)$ is overlapping with a white light supercontinuum of the fundamental $I_{WL}(\omega)$. The frequency beat between the white light and the SHG pulses separated by the time delay τ_0 and measured with the spectrometer can be written as [30]:

$$\begin{aligned} S(\omega) &= (1-a)I_{WL}(\omega) + aI_{SHG}(\omega) + 2\sqrt{a(1-a)I_{WL}(\omega)I_{SHG}(\omega)} \\ &\quad \cdot \cos(\phi_{SHG}(\omega) - \phi_{WL}(\omega) + \omega\tau_0 + \varphi) \end{aligned} \tag{3.4}$$

where a is the polarizer transmission for the SHG pulse and φ is the CE phase. If the number of spectral interferometric fringes is sufficient, the argument of the cosine in equation 3.4 can be resolved by the standard algorithm of Fourier-transform for spectral interferometry [32]. Basically, it consists in Fourier transforming the pattern into time domain, applying a bandpass filter around $t = \tau_0$ to select only the beating contribution, and doing the inverse Fourier-transform. The phase of the

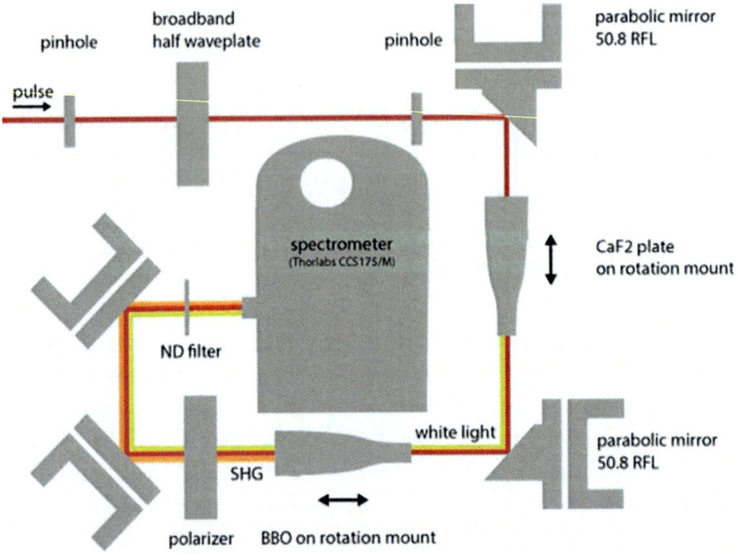

Fig. 3.7 f-2f interferometer scheme coupled to the filamentation broadening output to measure the CEP fluctuation and provide PID values to control the piezo mirror shifter installed in the OPA

retrieved complex spectral function contains the differential phase of the two fields $\text{mod}_{2\pi}\{\phi_{SHG}(\omega) - \phi_{WL}(\omega) + \varphi\}$. The relative variation of φ can then be extracted and is sent to the piezo amplifier by the mean of a digital PID controller. Note that such measurement doesn't allow to calculate the absolute value of φ as $\phi_{SHG}(\omega)$ and $\phi_{WL}(\omega)$ are not fully resolved.

When performing well, the active loop stabilization permits to reduce the measured free-running CEP of r.m.s. = 531 mrad jitter down to r.m.s. = 180 mrad and maintain it over 30 min as plotted in Fig. 3.8. Recently, with the idler shifted to 2 and 2.1 μm, the reflection of the filament cell Brewster window does not provide enough intensity to generate a stable white light continuum and longer f-2f signal integration time is required, reducing maximum feedback loop update frequency to <10 Hz. It should be replaced by the reflection of the uncoated fused silica GVD compensation plate or another dedicated reflection for better results.

3.3.4 Preparation of the Pump Pulses and Associated Specifications

Depending on the experiment and the measured phenomena, the requirement on the pump pulse might be different. As the concern of this work was to observe time resolved signals at the very limit of the actual SXR energy cutoff and over the full SXR water window, the results presented in the next chapters were all measured

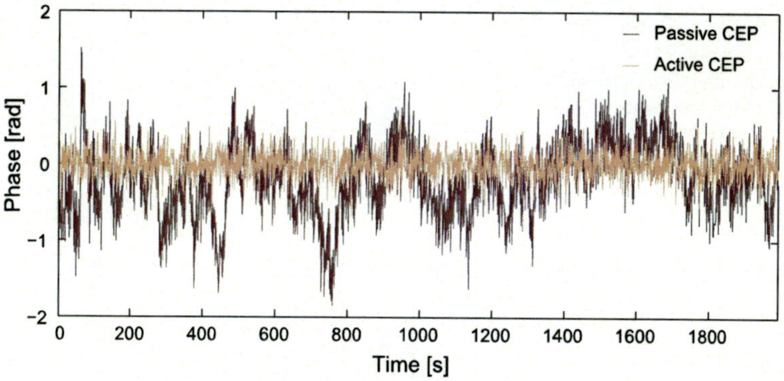

Fig. 3.8 1.8 μm idler CEP measured over 30 min using the f-2f interferometer. Passive free-running CEP (maple), corresponding to r.m.s. = 531 mrad jitter where the actively locked CEP (brown) stabilization reduces it to r.m.s. = 180 mrad. Reprinted with permission from [31] © The Optical Society

using NIR strong field ionization. In this regime, the amplitude of the electric field of the laser, $E_0 = \sqrt{2I/\epsilon_0 c}$, is of high importance as it has to be comparable to the Coulomb field (10^8–10^9 V/cm) between the electron and the binding ion-core of the measured molecule to initiate the ionization process. The fundamental wavelength of the pump pulse can be chosen between the 800 nm or the OPA signal. An attempt was done to extract the central fraction of the compressed 1.8 μm idler beam with a drilled-mirror but the remaining intensity to drive HHG in helium was too low to achieve enough SXR flux for the transient absorption measurements. Nevertheless, such configuration combined with lower IP gases such as neon or argon should be investigated in order to benefit of the sub 12 fs pulse duration resolution.

The 2 mJ of 800 nm separated from the OPA pump pulse with an 80:20 beam splitter was used for experiments on the gas phase while the OPA 1.3 μm signal wave was used to pump in liquids.

The delay line is realized with a 0.1 μm precision translation stage allowing to scan the delay between the pump pulse and the SXR probe with 0.2 μm, respectively 667 as time resolution. A shutter is set into the pump line and synchronized with the spectrum acquiring camera enabling the measurement of the pumped and unpumped absorption spectra at each delay step.

After about 7 m propagation to compensate for the OPA optical path length, the 800 nm pump pulse is measured with 50 fs and 1 mJ on target position. It is focused with a 40 cm focal length lens and the M2 coefficient is estimated of about 1.4. Assuming a Gaussian beam with a central hole of 3 mm due to the recombination mirror (see Sect. 3.4.2) leading to a Bessel type focus, the upper limit of the pump-pulse intensity at focus is estimated to be around $4 \times 10^{14} \frac{\mathrm{W}}{\mathrm{cm}^2}$, respectively an electric field amplitude of $1.7 \times 10^9 \mathrm{V/cm}$.

In the case of the 1.3 μm pulses, the temporal chirp and the spatial profile at the output of the OPA didn't allow for enough intensity at the target focal position requiring some more optical correction. The residual lateral divergence was first

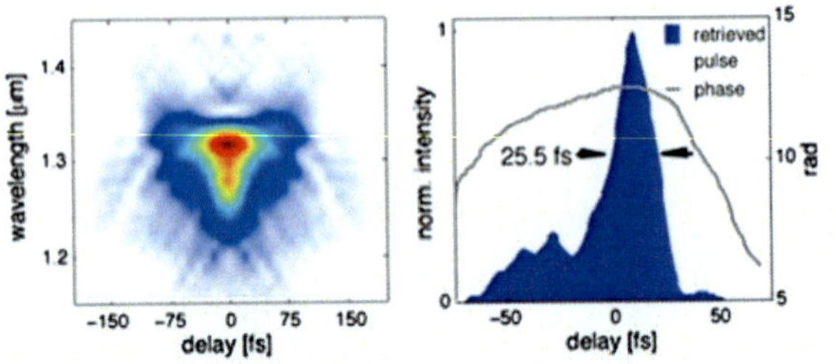

Fig. 3.9 SHG-FROG trace from the spatially filtered 1.3 μm pump pulse (left) and temporal intensity with phase (right) after chirp mirrors partial compression

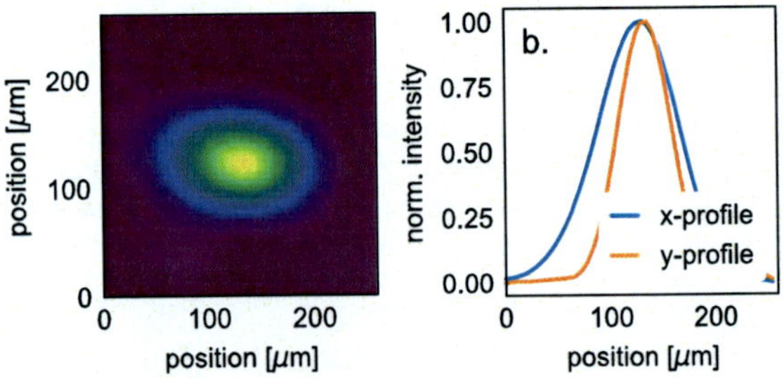

Fig. 3.10 1.3 μm pump pulse beam profile at the target position measured with a standard CCD sensor. The mean value of x- and y-axis FWHM is $100 \cdot \sqrt{2} \cong 142$ μm (where $\sqrt{2}$ corrects for the 2-photon absorption).

corrected by two cylindrical anti-reflective coated lenses in order to retrieve sufficient collimation. The spatial mode was then filtered by a 0.5 μm diameter pinhole installed at the 1 m focal position of a spherical mirror. To avoid filamentation and other nonlinear effects during propagation, the focusing line is inserted into a transparent vacuum cell of 1.5 m length pumped with a primary scroll pump. The residual distance required to compensate for the idler compression setup is realized with an extra round trip over two compressing mirrors introducing –135 fs^2 at 1.3 μm. The FROG trace and the temporal retrieved profile are given in Fig. 3.9.

After the delay stage, the pulse is focused by a Galilean telescope for increasing the numerical aperture with a focal length of 1.2 m. Considering a main pulse duration of 30 fs for half of the overall intensity, the measured energy of 0.7 mJ together with the 142 μm FWHM focal diameter shown at Fig. 3.10 leads to a pulse intensity of 5×10^{13}W/cm^2.

3.4 Vacuum Beam-Line System

Because of the argon, carbon, nitrogen and oxygen constitution of air which are exactly the main targeted element of most SXR studies and because the resulting mean attenuation length of SXR energies is very short (3 mm), the full SXR beam line, from generation to spectrum detection, has to propagate under vacuum. The total actual 3 m optical path is composed by a differentially pumped generation chamber, a pump-probe recombination and refocusing chamber, the interaction chamber and the SXR spectrometer. All elements are designed in order to support high vacuum pressure (10^{-6} mbar) conditions and such typical values are reach in every chamber when no experiment is running.

3.4.1 Soft X-Ray Generation

The SXR pulse is generated in neon by directly focusing the idler pulses with a CaF_2 400 mm focal length lens or in helium by focusing the compressed idler pulses using a 250 mm focal length spherical mirror into a static high-pressure gas target. The HHG chamber has been designed by Kristina Zinchenko with contributions of other members of the group of H.J. Wörner at ETHZ, and replicated for the experiments in Geneva by the ETHZ workshop. The focusing mirror is placed outside the chamber and the beam enters the vacuum setup through a CaF_2 window mounted at Brewster's angle. The generation high-pressure gas is confined in a lateral laser-drilled tube target of 6 mm or 4 mm diameter adjustable in XYZ by an external manual stage system. The interaction length can be roughly estimated to be about the same 6 mm or 4 mm. As shown in Fig. 3.11, the high-pressure stage, around the high-pressure target, is pumped by a 35 m^3/h scroll vacuum pump (by Edwards) and the two second differential pumping stages are connected to a 100 m^3/h industrial multi-roots pump (by Pfeiffer). A third differential pumping stage is pumped by a turbo-molecular vacuum pump in order to ensure good evacuation while HHG with high pressure and lightweight helium gas. Using this three-stage differential pumping vacuum design, the pressure is reduced down to 10^{-4} mbar range in the third stage of the chamber and to 10^{-6} mbar in the refocusing chamber.

When generating in neon, the maximal HHG flux is reached with the 6 mm diameter target and 700 mbar backing pressure. For these conditions, the driving

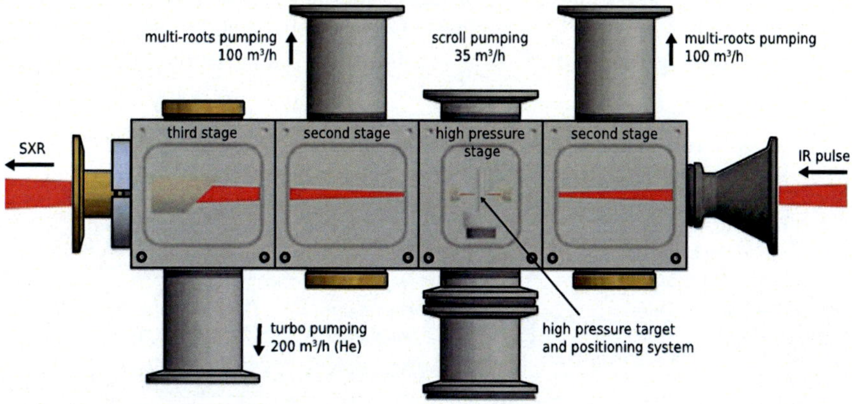

Fig. 3.11 High pressure differential pumping chamber for HHG generation. Each stage is isolated by minimum pinhole diameter along the beam axis to reduce gas flux and to filter the SWIR radiation after HHG. Design: Yoann Pertot and Mario Seiler, ETHZ

1.8 μm pulse was not compressed so that the 45 fs FWHM and 2 mJ idler energy leads to more than 0.5 PW/cm^2 at target position with a Rayleigh range of 15 mm. With these parameters it was possible to see a strong reddish plasma plume formation before and after the target. Depending on the experiment requirements, the HHG spectrum was shaped in order to obtain more flux in the higher (above 250 eV) or lower (below 250 eV) energy range; slightly increasing the backing pressures favored the emission of higher photon energies. The maximum measured cutoff generated in neon was 400 eV.

When generating in helium, in order to benefit from its higher ionization energy (24.6 eV) as compared to neon (21.6 eV), allowing for higher cut-off energy, the best generation conditions were found with the shorter interaction length of 4 mm and around 4 bar of backing pressure. In order to increase the peak intensity in the interaction volume, the post-compressed pulses are used, leading to 2.1 PW/cm^2 intensity and a Rayleigh range of 6 mm when focused with a $f = 250$ mm spherical mirror. The high-pumping capacity and the differential stage configuration permit reducing the pressure in the second stage to 10^{-2} mbar and to <10 mbar in the high-pressure stage, thus resulting in a strong helium pressure gradient. With this setup it was possible to increase the helium backing pressure to more than 6 bar without facing pumping saturation over time. Thus, the previously predicted [16] phase-matching pressure range between 3.5 and 5.5 bar can be covered for the 1.8 μm driver. As for the neon case, a strong plasma plume formation before and after the target is visible. The generated SXR beam is spatially separated from the co-propagating IR by 3 pinholes filtering the outer part of the highly divergent IR pulse.

In order to reach even higher cut-off energies and cover the oxygen k absorption edges, the idler wave of the OPA was recently tuned to a longer wavelength, around 2 and 2.1 μm, taking advantage of the $U_p \propto \lambda^2 \cdot I_S$ wavelength dependence. A comparison between the best conditions for SXR HHG spectra averaged over all CEP and generated at 1.8, 2 and 2.1 μm is shown in Fig. 3.12. In both cases of neon and helium generation target gas, the harmonic flux is optimized simultaneously for all photon energies by positioning the beam waist slightly after the center of the target. Such geometry induces a small intensity variation over the interaction

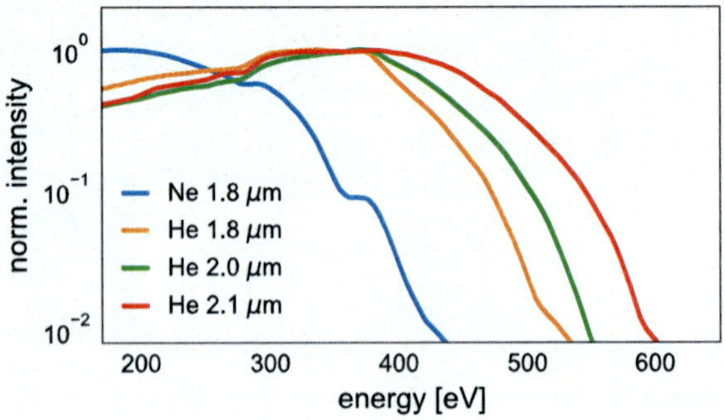

Fig. 3.12 SXR spectra, after Al filter, generated in neon (blue) and helium (others) for different driving pulse wavelength. In each case, generation parameters are optimized for maximum cutoff energy

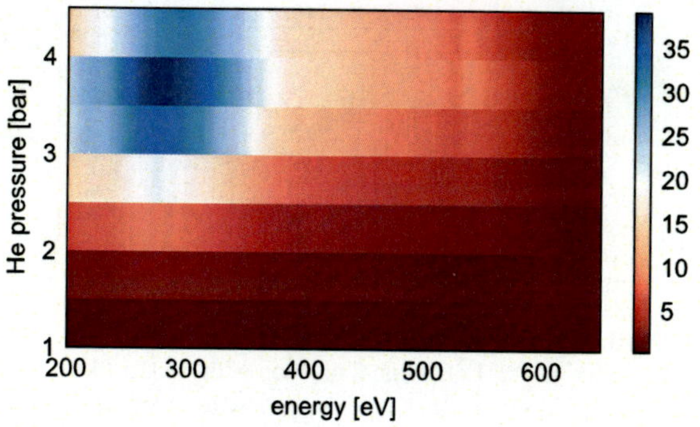

Fig. 3.13 Helium pressure generation map. The 2.1 μm pulse, 2.1 PW/cm^2 is keep with same parameters while only the gas backing pressure is adjusted. The spectra are recorded after an SiO_2 substrate to show the oxygen K-edge. Each spectrum is averaged over 3 min

length, reducing the intensity-dependent dipole phase gradient, which can explain the increase of harmonic flux observed for this relative focus position [33].

The relevant adjustable parameters for the optimization of the SXR signal are the pulse post compression and CEP, the pulse energy, the generation gas pressure and the intensity of the beam on the target. These parameters are either coupled together in some way or can be tuned independently. Yet, Fig. 3.13 shows a direct comparison of different helium pressures used to generate the harmonics. Highest flux and cutoff are observed for the same 4 bar of backing pressure. The 1.8 μm driving pulse is temporally characterized and discussed in the results chapter and the optimal (shortest) pulse parameters were always considered as a starting point before final optimization of the HHG generated spectrum.

The laser-drilled target hole diameter strongly affects the gas pressure distribution and therefore phase-matching conditions even though it is difficult to precisely characterize it. The target tube was usually replaced after about 20 h of operation because the HHG flux decreased or because the vacuum pumps could not cope with the increasing gas load. A large variability in both the hole diameters and shapes was observed after the 20 h operation. Typically, the output hole was about 3 times smaller than the input one. It is supposed that the evolution of the hole diameter during its duty lifetime strongly affects the pressure gradient, the effective intensity at focus position and the interaction length of the HHG. This parameter should be investigated more rigorously and simulated for assuring reproducible conditions over the experiments.

Under the conditions described above, efficient HHG is reached by strongly ionizing the generation gas (see the calculations of the ionization fractions shown by Fig. 3.14) and maybe creating a filament over the interaction length. Indeed, the competition between the generated plasma induced defocusing and the Kerr self-focusing effect might allow the pulse to significantly reshape in space and time over a few millimeters. Although the Rayleigh range (6 mm) is larger than the interaction length (4 mm) so that it is difficult to claim for a classical filament regime, the accumulation of nonlinear effects during the propagation leads to an important restructuring of the spatiotemporal pulse profile as the plasma refractive index is larger in the center of the pulse than on the edges. The trailing part of the pulse then

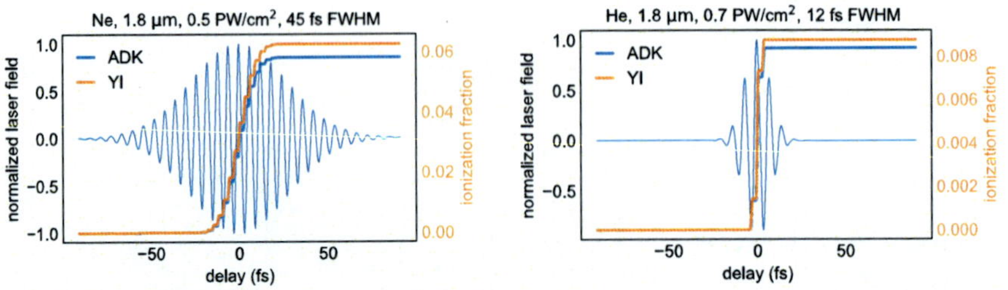

Fig. 3.14 Calculations for the ionization degrees of neon and helium with the ADK and YI models (see Sect. 2.2.5). Left: neon ionization with the 0.5 PW/cm^2 uncompressed 45 fs pulse. The HHG build up occurs in the central most intense 6 optical cycles. Right: helium ionization with the clamped 0.7 PW/cm^2 intensity of the 12 fs driver. The HHG build up occurs over 1.5 optical cycles

propagates faster than the center, inducing pulse steepening and self-compression [34]. Moreover, by calculating and measuring the ionization fraction of helium at 4 bar pressure and the resulting plasma defocusing for a 1.8 μm pulse, Johnson et al. shown that the intensity is clamped to about 0.7 PW/cm^2 over the propagation [33]. This effect explains the deviation between calculated and measured cut-off energies and would also contribute to extend the phase-matching range by reducing the plasma dispersion induced mismatch. To define this "overdriven HHG regime" [33], Johnson et al. use the ionization fraction F as a criterion: "the overdriven regime is entered for an 1800 nm pulse if the ionization defocusing in a thin medium clamps the peak intensity achieved to less than 70% of the vacuum propagation case". They provide an empirical formula to help the comparison between different setups. Based on the recent results of numerical simulations including all phase matching effects and showing that group velocity mismatch plays an important role for HHG with short SWIR driver pulses [35], this nonlinear pulse reshaping and intensity clamping supports a more complex approach to extend the coherent build-up range and generate bright SXR pulses [33, 36, 37].

The Fig. 3.15 displays the ionization fraction dependence on laser wavelength and pulse peak intensity of the HHG driving pulses in helium at three wavelengths

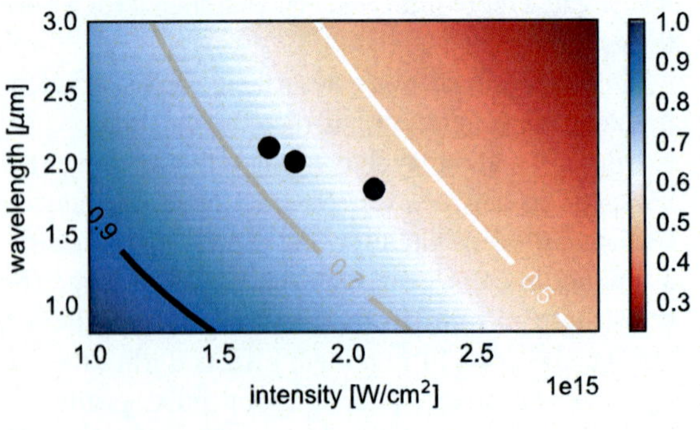

Fig. 3.15 Overdriving criterion [33] map together with the 3 pulse configurations used for HHG in helium. All of them are below the 0.7 criterion limit and are therefore in the "overdriving regime"

Table 3.2 Comparison of achieved SXR pulse energies over the water window between similar HHG setups

Ref	Gas	Interaction length mm	Pressure bar	Peak intensity PW/cm^2	Total WW energy pJ
This work	neon	6	0.7	0.5	–
Johnson et al. [33]	neon	0.8	1.1	> 2	71 (1 kHz)
Teichmann et al. [37]	neon	1.5	3.5	0.5	–
This work	helium	4	4	2.1	2 (1 kHz)
Johnson et al. [33]	helium	0.8	4	> 2	7.7 (1 kHz)
Teichmann et al. [37]	helium	1.5	12.5	0.5	0.9 (1 kHz)

(1.8, 2.0 and 2.1 μm). In all these cases, the suggested criterion for the overdriven regime is fulfilled.

Table 3.2 compares some generation parameters with similar setups from the literature. The resulting energy per pulse values over the water window are all in the same order of magnitude. The methods of flux determination are nevertheless different between studies; Johnson et al. [33] and Teichmann et al. [37] determine the harmonic flux from their photon-counting X-ray CCD camera, grating diffraction efficiency, spectrometer angular acceptance, filter transmission and the carbon line absorption. In contrast, the flux value determined for this setup results from a comparative flux calibration using a calibrated photodiode while keeping experimental conditions of alignment as similar as possible.

3.4.2 Spectral Filters and Refocusing Chamber

By entering the second vacuum chamber, the HHG radiation can be filtered from the remaining co-propagating SWIR pulses using specific energy metallic bandpass filters. The transmission curves in the spectral region of interest for the most used ones are represented in Fig. 3.16. 3 filters can be mounted together on a motorized wheel in order to select the appropriate one without venting the chamber.

The slightly divergent SXR beam is then refocused with a nickel-coated grazing-incidence toroidal mirror (by ARW Optical) onto the sample. Its 85 × 30 mm concave surface offers an f = 500 mm focal length at 3° grazing angle with <1 nm rms slope error. The radii of curvature of the mirror in horizontal $R_h = 19{,}107$ mm and vertical $R_v = 52.34$ mm directions satisfy the condition $R_v/R_h = \sin^2\alpha = 0.0027$ so that focal distances of the mirror are the same in the horizontal and vertical directions and the toroidal geometry allows one-to-one imaging of the point-like SXR source at distance r_1 (from the toroidal) onto the sample at distance r_2 (from the toroidal). In order to satisfy the Bragg condition over the entire area of the mirror, the SXR source

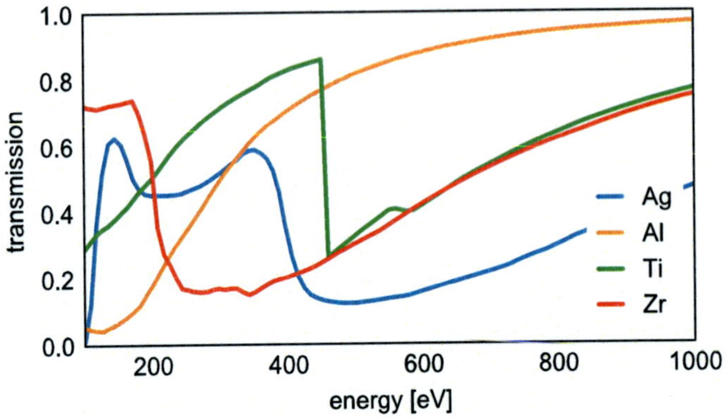

Fig. 3.16 Spectral transmission of the metallic filters used to remove SWIR radiation from SXR pulse. The filters can be remotely introduced into the beam before refocusing onto the sample and in the spectrometer. Because of its sharp $L_{2,3}$-edge, the Ti filter is also used for the spectrometer calibration. Data from [38]

and the sample are located on the Rowland circle in a symmetrical configuration, leading to the condition:

$$r_1 = r_2 = R_v \sin\alpha = \frac{R_h}{\sin\alpha} = 1m = 2f \tag{3.5}$$

The toroidal is mounted on a custom full piezo-motorized system allowing to adjust all Euler angles independently. A similar geometry and complete toroidal alignment discussion is available in the thesis of Lin [39].

After refocusing, the SXR pulse is recombined with the pump pulse (see Sect. 3.3.4) using a 45° annular mirror or a flat mirror. The annular mirror used to collinearly align the 800 nm beam has a central hole of 3 mm resulting in a Bessel type focus at sample position. To reflect maximum pulse energy, the flat mirror is used in case of pumping with the 1.3 μm signal. The latter leads to a non-collinear transient absorption geometry, where the angle between pump and probe is minimized to 2°.

3.4.3 Sample Interaction Chambers

The interaction chambers are different and dedicated to specific experiment. The gas phase measurement chamber is a 40 cm × 40 cm × 40 cm cube pumped by a 7200 m^3/h turbomolecular pump. It was brought to Geneva by the group of H.J. Wörner for this common experiment. The sample gas target for transient absorption measurements was very similar to the 6 mm HHG one. It consisted of a tube with two holes drilled by the 800 nm pump laser beam itself. The tube was placed inside a

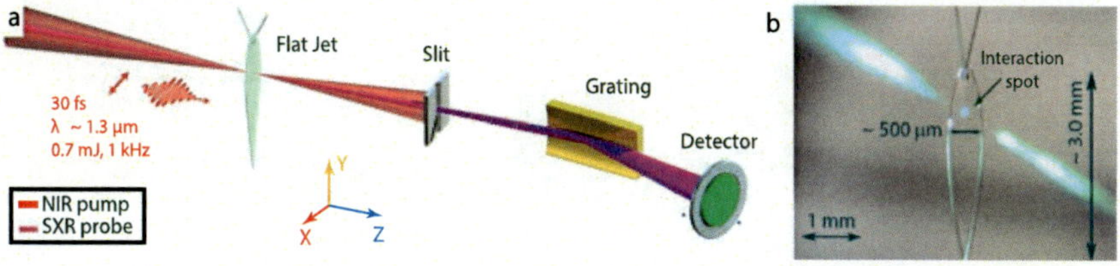

Fig. 3.17 Experimental scheme of TR-XAS in liquid samples. **a** The liquid flat microjet is positioned at the focal spatial overlap of the pump and the probe beam. **b** Photograph of the laser–liquid interaction. The sample is liquid ethanol. Adapted from [40], under CC-BY license

differentially-pumped cube located within the vacuum chambers. Backing pressures of 150 mbar for CF_4 and 20 mbar for SF_6 were applied to the target, resulting in background pressures in the low range of 10^{-4} mbar in the outer vacuum chambers.

The experiments in the liquid phase are performed using a flat jet system as shown in Fig. 3.17. The flat jet was developed at ETH and is comprised of two micro jets, taking inspiration from a commercially available system [41]. A simple explanation is that the collision of the two microjets results in the formation of several thin sheets, and finally the formation of a singular cylindrical jet. The liquid is pumped using a high-performance liquid chromatography (HPLC) pump selected for its high stability and capability to pump liquids with a consistent backing pressure at a desired flow rate. The overall setup can be used within a cubic vacuum chamber, in addition of two nitrogen cold traps to ensure suitable working range ($<5 \times 10^{-3}$ mbar). For the experiments presented within this thesis, 30 μm nozzles have been used to produce a flat micro jet of approximately 1 μm. The thickness of the jet depends on the flow rate, nature of the sample and nozzle size. The typical flow rate for stable operation with solvents was 1.3 ml/min. Under these conditions, the liquid exits the nozzles at about 10 m/s, which is fast enough to guarantee a complete recycling of the sample between successive measurement at 1 kHz repetition rate. More details on the setup are available in a recent publication from ETH [40], where T. T. Luu et al. used the same system to generate XUV pulses by HHG in liquids.

3.4.4 Pump-Probe Spatial and Temporal Overlap

The pump and probe spatial overlap is a redundant problematic occurring at every measurement alignment. Because the SXR beam is not always perfectly collinear with the remaining SWIR driver, overlapping the NIR pump and the SWIR driver as a proxy for the SXR beam is not enough. Typically, a thin aluminum foil is attached to the movable sample target in order to be at the exact same SXR-axis position that the center of the sample when in measurement position. The pump beam is then used to drill a small hole in the foil and the foil position is adjusted to maximize the SXR detected flux. The pump alignment is iteratively corrected until no adjustment of the

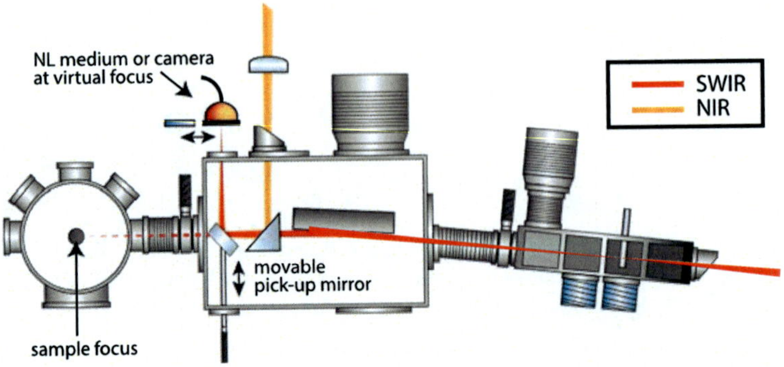

Fig. 3.18 Scheme of the pump-probe spatial and temporal overlap method. The pick-up mirror is used to monitor the beams after recombination. Design: Kristina Zinchenko, ETHZ

foil is required. Once satisfactory overlap is obtained, a camera is set at the virtual pump-probe focal position reached by a movable "pick-up" mirror allowing sending both beams outside the chamber after recombination. The positions of both spots on the camera are recorded and used as references for further alignment.

The determination of the pump-probe time overlap Δt_0 is done after the recombination mirror. The camera used as reference for the spatial overlap is replaced by an interaction medium (e.g. a BBO crystal) to create a nonlinear effect demonstrating the temporal overlap of the two pulses.

In the case of the 800 nm pump pulse, a fused-silica plate of 500 μm thickness is used as an interaction medium. The intense pump pulse transiently changes the refractive index of the thin fused silica plate, introducing a local increase of transmission for the 1800 nm pulse. For the 1.3 μm pump, the sum frequency generation $\omega_3 = \omega_1 + \omega_2$ occuring in a BBO crystal generates wavelengths around 800 nm observed directly on a screen or via a spectrometer. The delay introduced by the different group velocities of the HHG driver and the pump pulse in the CaF_2 chamber output window was considered. A schematic explanation of the method is shown in Fig. 3.18.

This method relies on the assumption that both driver and SXR maxima coincide. From the ionization degree calculations of neon and helium during HHG shown in Fig. 3.14, it is reasonable to consider that the high harmonics are mainly generated in the central part of the driving pulse in both cases. Hence, the peak intensities of the SWIR and the SXR pulses coincide, which assures precise determination of Δt_0.

3.5 SXR Spectrometer

The recent development of bright HHG X-ray sources can benefit from the previous and parallel achievements done on detectors by the third generation of synchrotron and free electron laser facilities. In the transient absorption experiment setup presented here, the SXR spectrometer is the final instrument and one of the most important. Its performance significantly affects the quality of the acquired data. In particular, the covered energy range, the efficiency and the resolving power are of most importance when designing or assembling the system for a specific experiment. Note that a flexible geometry and a standardized design are also relevant for such a research-oriented instrument.

In order to resolve the electronic transitions and energy values of the observed system, the probe beam spectrum is diffracted into a detector able to convert the energy into a digital value. Depending on the energy range, the approach to diffract the light can be different. The Bragg law can be used to resolve signals with crystal spectrometers in the hard X-ray regime [42, 43], while the required lattice constant of crystals is too high and the quality of the planes of these crystals not adapted for energy resolution in the soft X-rays. Pioneering works in the 80s [44, 45] demonstrated the possibility of using grating spectrometers based on the Rowland circle principle to perform high-resolution measurements in the soft X-ray regime, giving access to the study of core transition lines of systems containing vital elements like C, N, and O [46–53]. More recently, high resolution soft X-ray spectrometers based on varied line-spaced (VLS) gratings have been developed [48, 54–58]. Compared to the common grating spectrometer based on Rowland geometry, the VLS grating minimizes aberrations and focuses on a flat focal plane which facilitates the detection with X-ray CCDs [59] or array-type microchannel plates (MCP). The two-dimensional detector therefore becomes the energy resolution limiting factor. The distance between the grating and the detector is obviously also determinant. For instance, the actual SAXES spectrometer detector arm is 5 m long in order to increase the resolving power [56], giving the opportunity to directly observe vibrational states in the spectrum [60] of the ADRESS beam line [61] of Swiss Light Source of the Paul Scherrer Institute. Similar projects just implemented arm lengths up to 15 m to overpass the detector array resolution.

The purpose of this section is to present the different characteristics of the compact tabletop spectrometer versions used to collect the data presented in the result chapters. In total, three different instruments (G1, G2 and G3) were used depending on the campaign and the sample type (gas, solid, liquid). They are all based on a flat field spherical VLS grating and an array-type microchannel plate (MCP) with a phosphor screen imaged by a standard camera. Some details of each setup are provided by Table 3.3. For the recent campaign of liquid sample measurements, requiring

Table 3.3 Main characteristics of the custom SXR spectrometer configurations

Configuration	Grating model	r' mm	Array-type detector	Imaging camera	Pumping capacity M3/h
G1	30–001	235	MCP	ZWO ASI1600MM Pro	206–1240
G2	001–0659	563.2	MCP	PCO pco.edge 4.2	206–1240
G3	001–0659	563.2	X-ray CCD	–	

sustaining large differential pressures for good MCP operation, the geometry of the main chamber has been optimized.

3.5.1 SXR Flat Field Varied Line-Space Gratings

The energy diffraction optic is the fundamental element of the spectrometer. It defines most of the dimension of the chamber as well as the performance characteristics. At the soft X-ray domain, reflectivity of optics decreases very rapidly when the incident light angle (α in Fig. 3.19) becomes smaller, requiring adapted surface coating and small, grazing, angle for maximizing the efficiency. Concave gratings allow for systems free from focusing mirror designs. Different configurations for concave spectrometers have been designed. Some are variations of the Rowland circle geometry, while others, in combination with a VLS groove design, focus the spectrum on a flat field, which is suitable for coupling with an array-type detector. For a compact instrument compatible with the requirement of a tabletop soft X-ray HHG beam line, VLS concave gratings that focus onto a flat-field plane at a sub-meter distance are a sensible choice.

The VLS groove design allows the tangential focal curve to be shaped, while keeping the sagittal focal curve fixed. Such design allows compensating for the defocusing and aberrations introduced by the concave blank geometry on the diffracted energies, thus strongly increasing the resolving power of the system. The complete formalism of imaging properties for a concave VLS grating was developed [62] following the usual formalism for reflecting optics. The path function F given by Eq. 3.6 can be calculated and fully resolved for a given geometry.

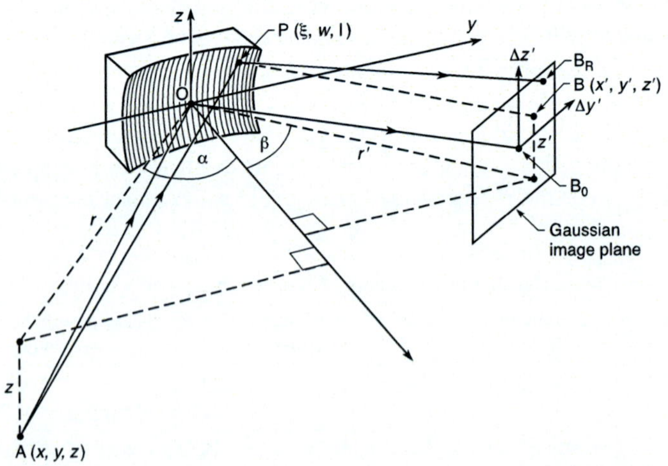

Fig. 3.19 Concave grating description coordinates. Reprinted with permission from [62] © The Optical Society

$$F = \sum_{ijk} F_{ijk} w^i l^j$$
$$\text{where } F_{ijk} = z^k C_{ijk}(\alpha, r) + z'^k C_{ijk}(\beta, r') + \frac{m\lambda}{d_0} f_{ijk} \quad (3.6)$$

with α and β the incident and reflected normal angle, r and r' the distance from the object to the grating center and from the grating to the detector, d_0 the initial groove separation, m the diffraction order and λ the diffracted wavelength. w and l being the grating surface coordinates as describes by Fig. 3.19. The C_{ijk} and f_{ijk} terms related to the strength of the i, j aberration of the wavefront and to the higher order groove pattern respectively are given up to the sixth order [63].

In a soft X-ray HHG beam line, the relatively low photon flux compared to synchrotron or free electron laser sources motivates to use the grating with the most efficient transmitting diffraction order m. Moreover, typical replica gratings for such application are gold coated and they are optimized for best performances at the 1st diffraction order. In the following, it will always be assumed that $m = 1$.

In the context of this work, we used two flat-field concave replica gratings of 2400 ln/mm groove density with different parameters. The two grating models are claimed by the manufacturers to be aberrations rectified and optimized for the same 206–1240 eV energy range. The typical parameters are defined in Fig. 3.20 and listed in Table 3.4.

The model 30–001 from Shimadzu relies on a holographic ruled VLS design whereas the 001–0659 from Hitachi [64] is mechanically ruled. The differences

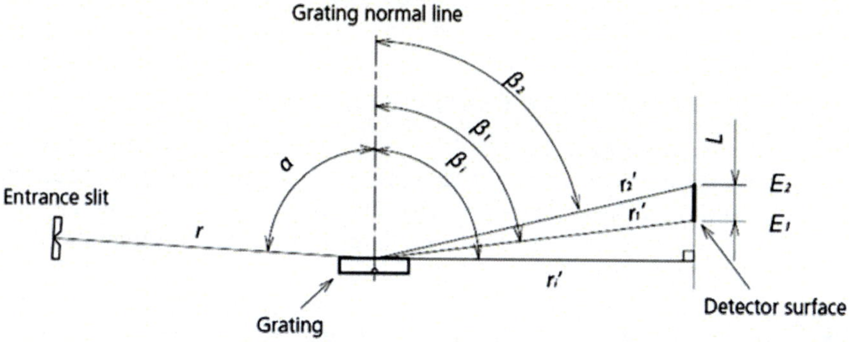

Fig. 3.20 Scheme of grating specifications

Table 3.4 Gratings manufacturer specifications

Model	Manufacturer	Grooves	Coating	E1–E2 eV	Plane length mm	R mm	α deg	β1 deg	β2 deg	ri mm	Bi deg	Size WxHxT
30–001	Shimadzu	Holographic	Au	206–1240	23.5	237.0	88.65	85.81	80.17	235.0	90	50 × 30x10
001–0659	Hitachi	Ruled	Au	206–1240	56.83	564	89	85.91	80.21	563.2		70 × 40x12

between holographic and ruled gratings are their groove profile and the manufacturing technique of their master pieces. Ruled gratings have a "sawtooth-shaped" groove profile where the blaze angle is calculated for maximum efficiency at a specific wavelength. The typical sinusoidal pattern of holographic surface is usually less efficient than a comparable ruled grating except when the groove spacing to wavelength ratio is nearly one [65]. Holographic replicated gratings generally have higher signal-to-noise ratio than those replicated from ruled master gratings, since they have no ghosts due to periodic errors in groove location and lower interorder stray light [65].

The 001–0659 model is designed for a larger incidence angle α and longer r and r' arms resulting in larger diffracted angles β and larger imaging focal plane. A similar replica grating is in use at the soft X-ray beamline P04 at PETRA III where the surface was inspected using an atomic force microscope. The measured slope error was 0.1 μrad rms [66]. The effective performance comparison between the two gratings is discussed in Sect. 3.5.5.

During measurements, both gratings suffered from carbon deposition on their surface. Such contamination deteriorates their performance and leads to a permanent absorption line on every measured spectrum. This occurred although the optics were kept under high vacuum chambers as some hydrocarbon and other gases (CH_4, CO, CO_2) present in the surrounding gases get dissociated by the high-intensity laser radiation over the surface of the grating. To get rid of this contamination, they were regularly exposed to uniform plasma discharges that sputters the carbon atoms from the surface [67]. Recently a strong UV lamp facing the grating was installed in the chamber in order to generate in-chamber ozone at every venting of the spectrometer [68].

3.5.2 Array-Type Micro Channel Plate Detector Stack

To convert the diffracted energy spectra into readable data, one needs a detector sensitive to the soft X-rays. Thanks to the flat-field imaging gratings, it is now possible to use array-type detectors capable of measuring a full or partial spectrum at the same time. Currently, most of the spectrometers are based on a charge-coupled device (CCD) or micro channel plates (MCP). Since the spatial resolution and efficiency of X-ray sensitive CCD can be higher than that of MCP [69], the large energy range of CCD make them ultra-sensitive to visible and infrared light.

Channel multipliers have a surface work functions which allow photoelectron production at incident wavelengths shorter than 200 nm. The BOS-40 beam observation system bought from Beam Imaging Solutions INC, rely on a 40 mm diameter active area dual MCP plates with a P-43 phosphor screen. The stack is assembled in a chevron configuration as shown by Fig. 3.21 within a 6 in. conflat flange with a glass view port. Each MCP has an active area composed of 10 μm diameter channels with 12 μm pitch and an 8° bias angle which gives a photoelectron conversion efficiency near 10% in the 170–800 eV region [70]. Such configuration should even reach 27% efficiency at 860 eV [71]. The P-43 phosphor screen is aluminum coated on the MCP

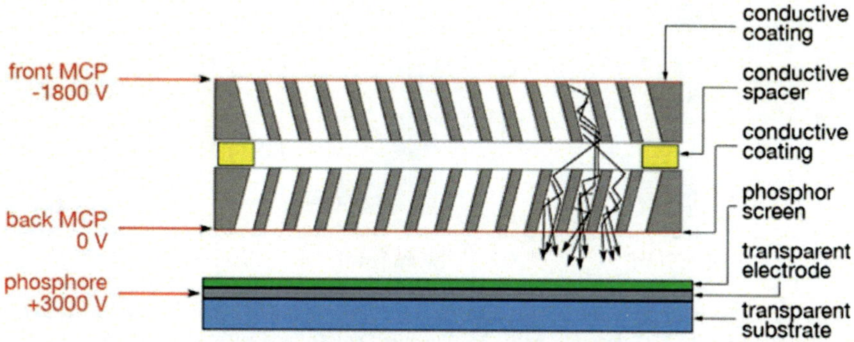

Fig. 3.21 Scheme of the MCP and phosphor plates assembly. The drawn arrows represent the electron cascading and spreading at inter-plates. Reprinted from [72], under CC-BY license

side to minimize for the fluorescence peak occurring at 545 nm to return toward the MCP and filter out the remaining contamination light.

To reduce ambient chamber electron and ion amplification by the MCP stack, a negative potential is applied to the front MCP while the back MCP is grounded [73]. The typical experimental voltages are –1800 V between front and back MCP and +3000 V between back MCP and phosphor plate. The maximum recommended potential across the MCP stack is –2000 V. Note that the conductive ring spacing the two MCP is at floating potential as the electrical impedance of both MCP is supposed to be identical. To operate safely and reduce the noise, the detection system requires high vacuum pressure (10^{-6} mbar) and the chamber ion-based gage is always switched off.

Such detectors are known to suffer from electron clouds spreading between MCPs and phosphor plate as represented in Fig. 3.21 [74–76]. While this electron distribution helps for the overall gain of the MCP stack, it decreases a lot the spatial resolution of the detector, thus reducing the spectral resolving power of the spectrometer. To quantify this spreading effect in experimental conditions, the phosphor screen emission corresponding to a single charged particle arriving into a micro channel was measured. The resulting spot sizes are given in Fig. 3.22 for different realistic MCP gains and are fitted with Gaussian functions to extract the FWHM mean value.

The mean spot size is 200 μm FWHM which is very large as compared to the initial 10 μm MCP channel diameter. It is, however, comparable to the value measured on similar systems [72, 76]. Given the Rayleigh criterion for instrumental resolution σ_R (i.e. the distance separating the two resolvable features): $\sigma_R \geq 0.849 \times FWHM =$ 170 mm. The respective effect on the effective spectral response of the spectrometer is discussed in detail in the Sect. 3.5.5.

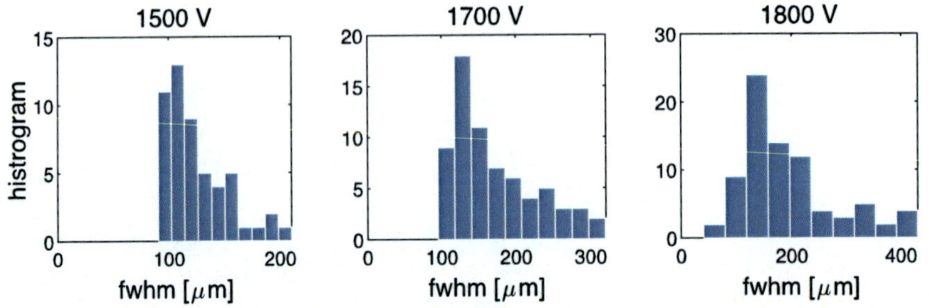

Fig. 3.22 Phosphor screen measured spot sizes induced by single photoelectron events at front MCP channel. The size depends on the MCP applied voltage and event intensity. At 1800 V, the mean FWHM spot size is 200 μm. The phosphor plate potential is 3000 V for all measurements

Table 3.5 Imaging camera technical data

Model	Sensor	Diagonal (mm)	Pixels	Pixel size (μm)	Exposure range (s)	Read noise	QE (at 545 nm) (%)	Dynamic range (bits)	Coupled objective NA
ZWO ASI1600MM Pro	CMOS	21.9	4656 × 3520	3.8	32 μs - 2000	1.2 e^-	60	12	1.4
PCO pco.edge 4.2	sCMOS	18.8	2048 × 2048	6.5	30 μs -10	1.3 e^-	68	16	0.95

3.5.3 Phosphor Plate Imaging Camera

To image the fluorescent phosphor plate and record the spectrum, a large numerical aperture objective is coupled to a standard CMOS camera at atmospheric pressure. The performance of CMOS cameras has made significant progress with very high detection efficiency at 545 nm, high dynamic range and high resolution. Moreover, the fast and low noise readout performances introduce no losses to the acquired data as they outreach the MCP stack resolution and signal-to-noise capacities. The Table 3.5 gives the main characteristics of the two-imaging camera used for measurements.

During the measurements, the energy spectrum region of interest was always centered on the camera in order to reduce Barrel distortion introduced by the large NA objective, and thus avoid energy resolution losses while converting 2D images to spectra by vertical integration. A typical experimental recorded image of the HHG spectrum is shown by Fig. 3.23.

3.5.4 SXR Spectrum Calibration

A calibration procedure is required to correct for grating misalignment and to find the absolute detector and imaging camera positions. It is necessary to repeat the calibration process every time the beam line has been modified or vented. In the case

Fig. 3.23 Experimental raw image of SXR spectrum generated in neon and acquired over 5 s with the pco.edge camera. The weak absorption at 285 eV originates from the grating carbon-contamination

of the results presented in this work, the calibration process was repeated for each set of data measurement. During measurements, a first order correction was directly applied to the spectrum in order to replace the pixel position by an explicit energy axis. All the final results have been post-calibrated for better accuracy.

The spectral calibration is based on known sharp absorption lines in addition to the carbon K-edge that is present in all spectra as a result of a residual carbon contamination on the surface of the grating (see Sect. 3.5.1). The typical lines used for calibration together with the sample type are listed in Table 3.6.

The sulfur, argon and N_2 known lines were only used in the case of gas and liquid phase experiments. In the following, a typical calibration of the spectrometer with the 001–0659 grating for the liquid alcohol ($C_nH_{2n+1}OH$) experiment is described.

In a first step, the background-subtracted transmitted spectra are overlapped together in order to extract the pixel value of the identified reference line as shown in Fig. 3.24.

The flat-field grating Eq. 3.9 [82] is used to fit the known energy points with the spectrometer geometry parameters. In the case presented here, the Ti at 454 eV, C at 285 eV and S at 196 eV, 184 eV and 173 eV reference lines were considered.

Table 3.6 Reference absorption lines used for calibration

Element	Assignment	Molecule	Phase	Energy eV	Ref.	Comment
Sulfur	$L_{2,3}$ ($2p_{3/2}$)	SF_6	Gas	173	[77]	
	$L_{2,3}$ ($2p_{1/2}$)	SF_6	Gas	184	[77]	
	$L_{2,3}$ ($2p_{1/2,3/2}$)	SF_6	Gas	196	[77]	
Argon	L_3 ($2p_{3/2}$)	Ar	Gas	249	[78]	
Carbon	K (1 s)	C	Solid	285	[79]	Grating contamination (REXAFs)
Nitrogen	K (1 s)	α-Si_3N_4	Solid	407	[80]	50 nm substrate
	K (1 s)	N_2	Gas	407	[80]	
Titanium	$L_{2,3}$ ($2p_{3/2}$)	Ti	Solid	454	[38]	Filter 100 nm
Oxygen	K (1 s)	SiO_2	Solid	534	[81]	100 nm substrate

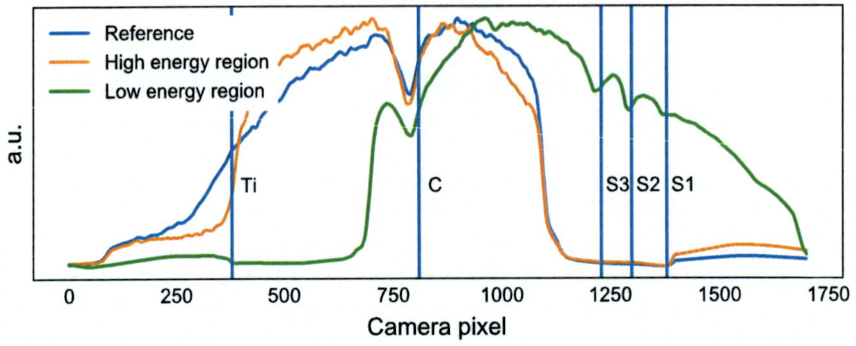

Fig. 3.24 Background-subtracted transmitted spectra with reference lines. Low pixel values correspond to higher photon energies. We clearly see the Ti $L_{2,3}$-edge, the C K-edge and the three S $L_{2,3}$-edges used for this calibration

$$\lambda = d_0 \cdot \left(\sin\alpha - \sin\left(\theta - \arctan\left(\cot\beta_0 + \frac{s(p-p_0)}{r'}\right)\right)\right)$$
$$\beta_0 = \arcsin\left(\sin\alpha - \frac{\lambda_0}{d_0}\right) \tag{3.7}$$

with d_0 the nominal groove spacing, s and p respectively the size and position of camera pixels, α the grating incident beam angle, θ the angle between grating and detector, r' the distance from grating to detector, λ_0 and p_0 the wavelength and pixel position of the known lines. The resulting energy calibration curve is shown in Fig. 3.25. Both given and retrieved fitting parameters are given in Table 3.7.

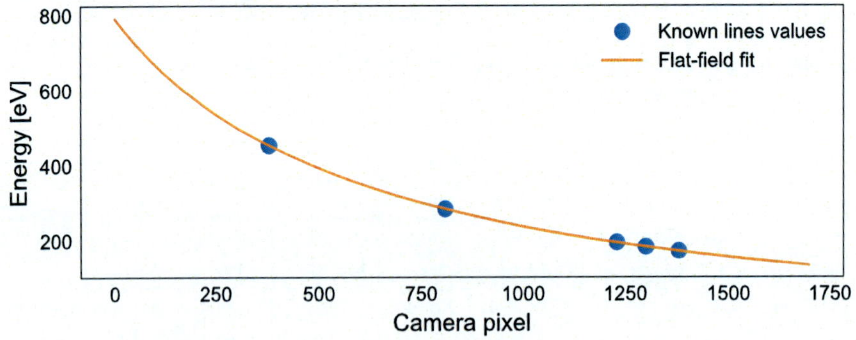

Fig. 3.25 Pixel to energy calibration curve fitting known lines with Eq. 3.7

Table 3.7 Typical flat-field equation fitting parameters. * The retrieved parameters

d_0 m	s m	p_0* pixel	α* deg	θ* deg	r' m
1/2,400,000	3.73e–05	4073	89.23	87.78	0.547

As the retrieved parameters are extremely sensitive to the pixel attribution of the reference lines, none of α^* nor θ^* retrieved parameters were considered to correct for the spectrometer misalignment. This should probably be investigated in the future. Nevertheless, the values were always contained in a $\pm$ 0.3° interval for the α and $\pm 1°$ interval for the θ as compared to the parameters provided by the manufacturer.

Once the energy calibration is applied, the grating-efficiency curve provided by the manufacturer is used to correct for the intensity. This intensity correction is only relevant for absolute emission and transmitted spectra as all molecular measured quantities are given in relative intensities. The correction is directly given by $I_{\text{corr}} = e_{\text{grating}} \cdot I_{\text{measured}}$ as the MCP, phosphor plate and camera sensitivity are expected to be constant over this energy range. To represent the spectra on an energy scale, the Jacobian conversion [83] is applied to the data. Indeed, the energy conservation involves Eq. 3.8:

$$f(E) = f(\lambda)\frac{\mathrm{d}\lambda}{\mathrm{d}E} = f(\lambda)\frac{\mathrm{d}}{\mathrm{d}E}\left(\frac{\hbar c}{E}\right) = -f(\lambda)\left(\frac{\hbar c}{E^2}\right) \tag{3.8}$$

After final normalization on the maximum intensity value, a typical generated spectrum together with the reference lines is displayed in Fig. 5.9.

3.5.5 *Efficiency and Spectral Performances*

The performances of a spectrometer typically rely on its spectral efficiency and resolving power. Since the overall efficiency response of the devices has never been measured experimentally on this setup, it can only be estimated from the grating and MCP stack efficiency curves given by the manufacturer with $e_{spectro} = e_{grating} e_{MCP}$. The grating efficiency curve is shown in Fig. 3.26.

These efficiency values do not take into account the spectrometer entrance slit size used in the spectrometer configuration G2. The efficiency response of the instrument

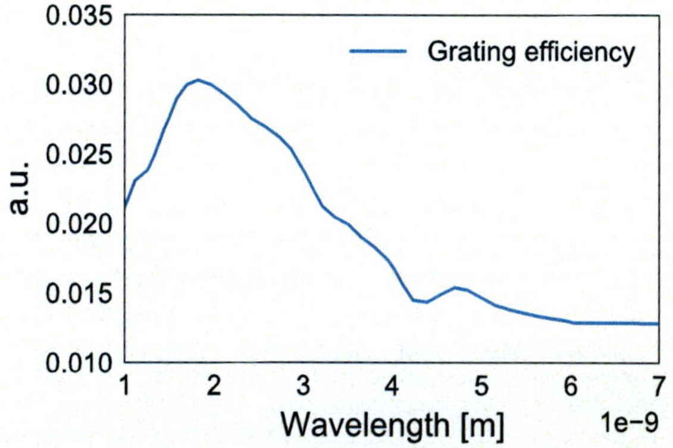

Fig. 3.26 30-01 grating efficiency curve provided by the manufacturer

is important as it directly impacts the signal-to-noise ratio and acquisition time. At the soft X-ray energies, it is difficult to really increase the reflectivity of the grating at 1st order and the two grating models described in Sect. 3.5.1 are both first generation of replicas. Note that the typical 10% MCP detector efficiency can be almost doubled by the use of CsI coating on the front MCP surface and a repeller grid in front of the MCP [73]. The CsI coating increases the photoelectron effect where the repeller grid set at front MCP potential avoid the highly negative potential of the front MCP to repel photoelectrons generated at its surface or at the very entrance of a channel.

One important characteristic of an optical element is its capacity to show separate images of very closely placed objects. It is called the resolving power R and can be derived from the well-known Rayleigh criterion. In the case of energy diffraction, the criterion gives: $\max\lambda' = \min\lambda$ and defines the ability of a system to form separate diffraction maxima of two closely separated wavelengths. By definition, $R = \frac{\lambda}{\Delta\lambda} = \frac{E}{\Delta E}$ and from the grating theory, $R = mN = N$ where N is the number of grooves illuminated. This value is the theoretical maximal achievable resolving power given by the grating only. It does not consider its surface geometry nor the spectrometer design and thus does not define the performance of the final spectrometer system.

The effective instrumental line spread function can be estimated by the vector sum of the main contributions to the resolution derived from Eq. 3.6 [62]:

$$\Delta E_{eff} = \sqrt{\Delta E_{slit}{}^2 + \Delta E_{det}{}^2 + \Delta E_A{}^2 + \Delta E_{se}{}^2} \tag{3.9}$$

where $\Delta E_{slit} = \frac{Sd_0\cos\alpha}{mrK}E^2$ gives the slit size S contribution, $\Delta E_{det} = \frac{\sigma d_0\cos\beta}{mrK}E^2$ being the effect of the limited spatial resolution of the array detector σ along the diffraction axis, $\Delta E_A = \frac{d_0}{m}\left(\frac{\partial F}{\partial w}\right)$ is the aberration term derived from Eq. 3.6 and $\Delta E_{se} = \frac{2E}{\tan\alpha - \beta}\Delta\phi$ the contribution from the slope error $\Delta\phi$ induced by micro-roughness imperfection on the grooves. The $K = 1.24 \times 10^6$ constant is to convert from wavelengths to energy units. As the two used gratings are presented to be Coma and aberration corrected by the manufacturer and also because the F_{ijk} coefficients are not fully known for these two gratings, the ΔE_A term is discarded. The $\Delta\phi$ values considered in the analytical resolution estimations are 2.2 μrad rms for the 001-0659 grating model [66] and 1.6 μrad rms for the 30-001 model (from the manufacturer). Considering the measured MCP detector spatial resolution of $\sigma =$ 170mm, as described in Sect. 3.5.2, the spectral resolution and resolving power for a 250 μm slit size, corresponding to the measured SXR beam size at target position, are represented in Fig. 3.27.

The MCP stack is by far the resolution limiting factor in the two designs. In such case, the bigger the diffraction, the higher the resolution as it helps to overreach the spatial resolution. Therefore, the G2 design with the 001-0659 grating offers a resolution of 1.2 eV at the carbon K-edge (285 eV) and 3.2 eV at the oxygen K-edge (534 eV) where the G1 instrument only resolve 3 eV at carbon and 7.9 eV at oxygen K-edges.

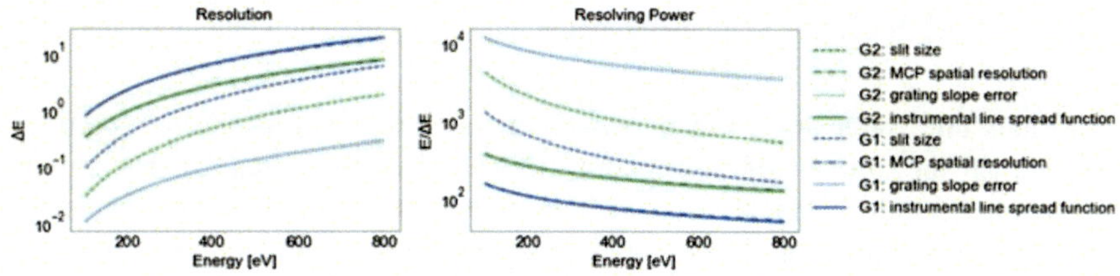

Fig. 3.27 Resolution and resolving power calculated for G1 (blue) and G2 (green) for the main contributions. The instrumental line spread function is the vector sum of each contribution

These bad resolving power due to the MCP electron cloud (see Sect. 3.5.2) is, of course, not adequate for tracking small spectral shifts at the X-ray absorption near edge structure or oscillations at extended X-ray absorption fine structure. By replacing the MCP detector with a 13 μm pixel size X-ray CCD camera, the G3 geometry should achieve the resolutions of 0.3 eV and 0.9 eV at carbon and oxygen K-edges respectively. The recent calibration of the G3 spectrometer using the N_2 gas 1 s to $1\pi_g$* transition demonstrated the ability to resolve 1 eV at the nitrogen K-edge (405 eV), assuming that the natural linewidth is 0.5 eV. Such configuration (i.e. with X-ray CCD detector) becomes limited by the entrance slit size, respectively the imaging spot size, and an even better resolution should be reached by further improvements of the SXR focal spot at the sample position.

3.5.6 *Transient Absorption Signal Acquisition*

Once pump-probe spatial and time overlapping is achieved (see Sect. 3.4.3), a background is recorded with similar acquisition settings and ambient light condition as during the experiment. The MCP voltage is reduced to avoid random noise from amplified photoelectron events at the MCP front surface. The scan is then automatically driven by a LabView software that synchronizes the camera capture frame rate with the pump line shutter and the delay stage. For each time step, a series of about 100 frames is recorded with the shutter alternatively opened and closed, leading to 50 pump-on and pump-off pair images. Every single pass scan is reproduced multiple times in order to remove artifacts from SXR spectrum shape or from flux variation and other environmental fluctuations.

Reference spectra with no sample as well as spatial overlap at virtual focal position (on the reference camera, see Sect. 3.4.3) were additionally recorded and compared before and after each scan to check the SXR beam stability and pump alignment. Since the time scale for a scan is only about 45–60 min, only small drifts were observed. However, if reference spectra before and after the scan are different, the data can be discarded.

3.6 XANES Data Normalization

The measured intensities, when performing a transient absorption experiment, depend on the amount of information desired. Assuming that the background, accounting for the residual ambient light and the noise level of the detector, is directly subtracted from all measured spectral intensities, the required quantities, as a function of the time delay between the pump and the probe t are: $I_0(t)$, that is the spectrum transmitted through the neutral sample (i.e. with pump off), and $I(t)$ which is the spectrum transmitted through the excited sample (i.e. with pump on). Recording $I_0(t)$ at each time step allows correcting for the laser intensity and sample interaction fluctuations. Nevertheless, if the experiment does not allow recording $I_0(t)$ for each time step, it can be assumed to be constant or equivalent to $I_0(t < 0)$ (where $t = 0$ is the overlap between the pump and the probe pulses, and negative time delays correspond to the probe preceding the pump). From these two quantities, the transient signal $\Delta A(t)$, given in ΔOD, is defined by:

$$\Delta A(t) = A(t) - A_0(t) = \ln\left(\frac{I_{\text{ref}}}{I(t)}\right) - \ln\left(\frac{I_{\text{ref}}}{I_0(t)}\right) = \ln\left(\frac{I_0(t)}{I(t)}\right) \tag{3.10}$$

where I_{ref} represents the probe spectrum without sample interaction. If this value cancels while calculating this transient signal quantity, it can be of great interest to represent the absorbance $A(t)$ and $A_0(t)$ separately and compare it with the literature and theoretical calculations.

Moreover, if the sample is only partially ionized, a fraction of the neutral absorbance can be subtracted from $A_0(t)$. The absorbance quantity of the excited sample thus becomes:

$$A(t) = \ln\left(\frac{I_{\text{ref}}}{I(t)}\right) - \beta\ln\left(\frac{I_{\text{ref}}}{I_0(t)}\right) \tag{3.11}$$

where β is a real number between 0 and 1 describing the non-ionized fraction of the probed sample. The motivation to subtract a non-ionized fraction can have two reasons. A certain fraction of the sampled molecules may remain neutral in the interaction volume of the pump beam and the excited and probed volumes may not overlap perfectly. The value of β is then estimated by comparison between calculated and observed absorption spectra.

With the current SXR flux in the water window and the sample availability (i.e. non-recycled gas and liquid samples) requiring short measurement times at each step, the signal-to-noise ratio in the EXAFS region of the spectrum is too weak so that only the XANES information can be extracted. In the following, the typical procedure for data normalization of XANES signal is detailed for the carbon K-edge of CF_4.

Typical near-edge absorption spectra of CF_4 before and after strong-field ionization and subsequent dissociation are shown in Fig. 3.28. The aim of the data normal-

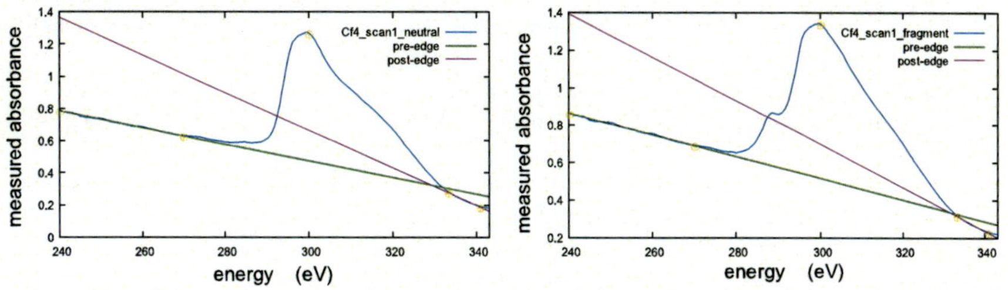

Fig. 3.28 Pre-edge and post-edge regions of a typical carbon K-edge absorption spectrum fitted by linear functions before (left) and after (right) strong-field ionization. The green fit of the pre-edge region is subtracted for analysis. The data are normalized on the post-edge function

ization is to get rid of the constant background and measurement artifacts and extract standardized data for further peak fitting and cross-setup comparison. As the measured spectra cover the XANES and part of the EXAFS regions that provides sufficient homogeneity for linear background subtraction, the pre-edge and post-edge regions are fitted with linear functions as shown in Fig. 3.28. The data are then subtracted from the pre-edge function and normalized to 1 on the post-edge function.

For time-scan data normalization, the same functions are applied to the entire scan after verification of their equality, as the objective is to normalize over a constant background. In the present case, the pre-edge function is $f(x) = 1.24 - 0.0032x$ and the post-edge function is $f(x) = 5.33 - 0.0155x$ (where x is in eV). They are both satisfying for neutral and excited CF_4 as shown in Fig. 3.28.

References

1. Goulielmakis E et al (2010) Real-time observation of valence electron motion. Nature 466(7307):739–743
2. Holler M, Schapper F, Gallmann L, Keller U (2011) Attosecond electron wave-packet interference observed by transient absorption. Phys Rev Lett 106(12):123601
3. Schultze M et al (2014) Attosecond band-gap dynamics in silicon. Science (80) 346(6215):1348–1352
4. Jordan I, Huppert M, Brown MA, van Bokhoven JA, Wörner HJ (2015) Photoelectron spectrometer for attosecond spectroscopy of liquids and gases. Rev Sci Instrum 86(12):123905
5. Pertot Y et al (2017) Time-resolved x-ray absorption spectroscopy with a water window high-harmonic source. Science (80) 355(6322):264–267
6. Saito N, Ishii N, Kanai T, Watanabe S, Itatani J (2016) Attosecond streaking measurement of extreme ultraviolet pulses using a long-wavelength electric field. Sci Rep 6(1):35594
7. Cousin SL et al (2017) Attosecond streaking in the water window: a new regime of attosecond pulse characterization. arXiv:1710.02208
8. Li J et al (2017) 53-attosecond X-ray pulses reach the carbon K-edge. Nat Commun 8(1):186
9. Gaumnitz T et al (2017) Streaking of 43-attosecond soft-X-ray pulses generated by a passively CEP-stable mid-infrared driver. Opt Express 25(22):27506–27518
10. Couairon A, Mysyrowicz A (2007) Femtosecond filamentation in transparent media. Phys Rep 441(2):47–189

11. Bergé L, Skupin S, Nuter R, Kasparian J, Wolf J-P (2007) Ultrashort filaments of light in weakly ionized, optically transparent media. Rep Prog Phys 70(10):1633
12. Ishii N, Kaneshima K, Kitano K, Kanai T, Watanabe S, Itatani J (2014) Carrier-envelope phase-dependent high harmonic generation in the water window using few-cycle infrared pulses. Nat Commun 5:3331
13. Gruson V et al (2017) 2.5 TW, two-cycle IR laser pulses via frequency domain optical parametric amplification. Opt Express 25(22):27706–27714
14. Ciriolo A et al (2017) Optical parametric amplification techniques for the generation of high-energy few-optical-cycles IR pulses for strong field applications. Appl Sci 7(3):265
15. Popmintchev T et al (2012) Bright coherent ultrahigh harmonics in the keV X-ray Regime from mid-infrared femtosecond lasers. Science (80), 336(6086):1287–1291
16. Popmintchev T et al (2009) Phase matching of high harmonic generation in the soft and hard X-ray regions of the spectrum. Proc Natl Acad Sci 106(26):10516–10521
17. Krause JL (1992) High-order harmonic generation from atoms and ions in the high intensity regime. Phys Rev Lett 68(24):3535–3538
18. Tate J, Auguste T, Muller HG, Salières P, Agostini P, DiMauro LF (2007) Scaling of wave-packet dynamics in an intense midinfrared field. Phys Rev Lett 98(1):13901
19. Béjot P, Hertz E, Kasparian J, Lavorel B, Wolf J-P, Faucher O (2011) Transition from plasma-driven to Kerr-driven laser filamentation. Phys Rev Lett 106(24):243902
20. Hauri CP et al (2004) Generation of intense, carrier-envelope phase-locked few-cycle laser pulses through filamentation. Appl Phys B 79(6):673–677
21. Schmidt BE et al (2014) Intense few-cycle infrared laser pulses at the advanced laser light source. Chin J Phys 52(1):537–545
22. Agrawal G (2001) Nonlinear fiber optics. Elsevier Science
23. Milián C et al (2015) Laser beam self-symmetrization in air in the multifilamentation regime. J Phys B At Mol Opt Phys 48(9):94013
24. Prade B et al (2006) Spatial mode cleaning by femtosecond filamentation in air. Opt Lett 31(17):2601–2603
25. Liu W, Chin SL (2007) Abnormal wavelength dependence of the self-cleaning phenomenon during femtosecond-laser-pulse filamentation. Phys Rev A 76(1): 013826
26. Tsang T, Krumbügel MA, Delong KW, Fittinghoff DN, Trebino R (1996) Frequency-resolved optical-gating measurements of ultrashort pulses using surface third-harmonic generation. Opt Lett 21(17):1381–1383
27. Trebino R et al (1997) Measuring ultrashort laser pulses in the time-frequency domain using frequency-resolved optical gating. Rev Sci Instrum 68(9):3277–3295
28. Sansone G et al (2006) Isolated single-cycle attosecond pulses. Science (80) 314(5798):443–446
29. Silva F, Teichmann SM, Cousin SL, Hemmer M, Biegert J (2015) Spatiotemporal isolation of attosecond soft X-ray pulses in the water window. Nat Commun 6:6611
30. Baltuska A et al (2003) Phase-controlled amplification of few-cycle laser pulses. IEEE J Sel Top Quantum Electron 9(4):972–989
31. Schmidt C et al (2018) High-order harmonic source spanning up to the oxygen K-edge based on filamentation pulse compression. Opt Express 26(9):11834
32. Takeda M, Ina H, Kobayashi S (1992) Fourier-transform method of fringe-pattern analysis for computer-based topography and interferometry. J Opt Soc Am 72(1):156
33. Johnson AS et al (2018) High-flux soft x-ray harmonic generation from ionization-shaped few-cycle laser pulses. Sci Adv 4(5):eaar3761
34. Panagiotopoulos P, Whalen P, Kolesik M, Moloney JV (2015) Super high power mid-infrared femtosecond light bullet. Nat Photon 9(8):543–548
35. Hernández-García C et al (2016) Group velocity matching in high-order harmonic generation driven by mid-infrared lasers. New J Phys 18(7):073031
36. Kleine C et al (2019) Soft X-ray absorption spectroscopy of aqueous solutions using a table-top femtosecond soft X-ray source. J Phys Chem Lett 10(1):52–58

37. Teichmann SM, Silva F, Cousin SL, Hemmer M, Bieger J, 0.5-keV Soft X-ray attosecond continua Nat Commun 7(11493 EP) 2016
38. Henke BL, Gullikson LM, Davis JC (1993) X-ray interactions: photoabsorption, scattering, transmission, and reflection at E = 50-30,000 eV, Z = 1-92. Data Nucl Data Tables 54(2):181–342
39. Lin M-F (2013) Ultrafast chemical dynamics of atoms and molecules studied by femtosecond time-resolved extreme ultraviolet transient absorption spectroscopy. UC Berkeley
40. Luu TL et al (2018) Extreme–ultraviolet high–harmonic generation in liquids. Nat Commun 9(1):3723
41. Ekimova M, Quevedo W, Faubel M, Wernet P, E. TJ Nibbering (2015)A liquid flatjet system for solution phase soft-x-ray spectroscopy. Struct Dyn 2(5):054301
42. Johann HH (1931) Die erzeugung lichtstarker röntgenspektren mit hilfe von konkavkristallen. Zeitschrift für Phys 69(3–4):185–206, May 1931
43. Hámos LV (1932) Röntgenspektroskopie und abbildung mittels gekrümmter kristallreflektoren. Naturwissenschaften 20(38):705–706
44. Callcott TA, Tsang K, Zhang CH, Ederer DL, Arakawa ET (1986) High-efficiency soft x-ray emission spectrometer for use with synchrotron radiation excitation. Rev Sci Instrum. 57(11):2680–2690
45. Nordgren J, Bray G, Cramm S, Nyholm R, Rubensson JE, Wassdahl N (1989) Soft x-ray emission spectroscopy using monochromatized synchrotron radiation (invited). Rev Sci Instrum 60(7):1690–1696
46. Yin Z et al (2014) Probing the hofmeister effect with ultrafast core–hole spectroscopy. J Phys Chem B 118(31):9398–9403
47. Schreck S et al (2014) Reabsorption of soft X-ray emission at high X-ray free-electron laser fluences. Phys Rev Lett 113(15):153002
48. Harada Y et al (2012) Ultrahigh resolution soft x-ray emission spectrometer at BL07LSU in SPring-8. Rev Sci Instrum 83(1): 013116 2012
49. Weinhardt L et al (2013) RIXS investigations of liquids, solutions, and liquid/solid interfaces. J Electron Spect Relat Phenom 188:111–120
50. Rubensson J-E, Hennies F, Pietzsch A (2013) High-resolution resonant inelastic soft X-ray scattering applied to liquids. J Electron Spect Relat Phenom 188:79–83
51. Beye M et al (2013) Stimulated X-ray emission for materials science. Nature 501(7466):191–194
52. Aziz EF (2010) The solvation of ions and molecules probed via soft X-ray spectroscopies. J Electron Spect Relat Phenom 177(2–3):168–180
53. Mitzner R et al (2013) L-edge X-ray absorption spectroscopy of dilute systems relevant to metalloproteins using an X-ray free-electron laser. J Phys Chem Lett 4(21):3641–3647
54. Dallera C et al (1996) Soft X-ray emission spectroscopy at ESRF beamline 26 based on a helical undulator. J Synchrotron Radiat 3(5): 231–238
55. Chiuzbăian SG et al (2014) Design and performance of AERHA, a high acceptance high resolution soft x-ray spectrometer. Rev Sci Instrum 85(4):043108
56. Ghiringhelli G et al (2006) SAXES, a high resolution spectrometer for resonant x-ray emission in the 400–1600eV energy range. Rev Sci Instrum 77(11):113108
57. Chuang Y-D et al (2005) High-resolution soft X-ray emission spectrograph at advanced light source. J Phys Chem Solids 66(12):2173–2178
58. Fuchs O et al (2009) High-resolution, high-transmission soft x-ray spectrometer for the study of biological samples. Rev Sci Instrum 80(6):063103
59. Dinardo ME et al (2007) Gaining efficiency and resolution in soft X-ray emission spectrometers thanks to directly illuminated CCD detectors. Nucl Instrum Methods Phys Res Sect A Accel Spect, Detect Assoc Equip 570(1):176–181
60. Hennies F et al (2010) Resonant inelastic scattering spectra of free molecules with vibrational resolution. Phys Rev Lett 104(19):193002
61. Strocov VN et al High-resolution soft X-ray beamline ADRESS at the swiss light source for resonant inelastic X-ray scattering and angle-resolved photoelectron spectroscopies. J Synchrotron Radiat 17(5):631–643

62. Noda H, Namioka T, Seya M (1974) Geometric theory of the grating J Opt Soc Am 64(8):1031
63. Chrisp MP (1983) Aberrations of holographic toroidal grating systems. Appl Opt 22(10):1508
64. Harada T, Takahashi K, Sakuma H, Osyczka A (1999) Optimum design of a grazing-incidence flat-field spectrograph with a spherical varied-line-space grating. Appl Opt 38(13):2743
65. Palmer CH (2005) Diffraction grating handbook, vol 1, no 6
66. Yin Z et al (2015) A new compact soft x-ray spectrometer for resonant inelastic x-ray scattering studies at PETRA III. Rev Sci Instrum 86(9):093109
67. Eggenstein F, Senf F, Zeschke T, Gudat W (2001) Cleaning of contaminated XUV-optics at BESSY II. Nucl Instrum Methods Phys Res Sect A Accel Spect, Detect Assoc Equip 467–468:325–328
68. Watts B, Pilet N, Sarafimov B, Witte K, Raabe J (2018) Controlling optics contamination at the PolLux STXM. J Instrum 13(4):C04001–C04001
69. Skinner CH, Schwob JL (1996) Charge-coupled-device detection of soft x rays for grazing-incidence spectrometers. Appl Opt 35(22):4321
70. Smith DG, Pounds KA (1968) Soft X-ray photon detection and image dissection using channel multipliers. IEEE Trans Nucl Sci 15(3):541–550
71. Bjorkholm PJ, Van Speybroeck LP, Hecht M (1977) X-ray quantum efficiency of microchannel plates. Int Soc Opt Photon 0106:189–195
72. Bolton PR et al (2014) Instrumentation for diagnostics and control of laser-accelerated proton (ion) beams. Phys Medica 30(3):255–270
73. Pearson JF (1984) Advances in soft x-ray performance of microchannel plate detectors. Phys Astron
74. Hoenderken TH, Hagen CW, Barth JE, Kruit P, Nützel GO (2001) Influence of the microchannel plate and anode gap parameters on the spatial resolution of an image intensifier. J Vac Sci Technol B Microelectron Nanom Struct 19(3):843
75. Lupone S, Soulisse P, Roncin P (2018) A large area high resolution imaging detector for fast atom diffraction. Nucl Instrum Methods Phys Res Sect B Beam Interact Mater Atoms 427:95–99
76. Wang X, Chini M, Cheng Y, Wu Y, Chang Z (2013) In situ calibration of an extreme ultraviolet spectrometer for attosecond transient absorption experiments. Appl Opt 52(3):323
77. Hudson E et al (1993) High-resolution measurements of near-edge resonances in the core-level photoionization spectra of SF6. Phys Rev A At Mol Opt Phys 47(1):361–373
78. Nakamura M et al (1968) Absorption structure near the LII, III edge of argon gas. Phys Rev Lett 21(18):1303–1305
79. Denley D, Perfetti P, Williams RS, Shirley DA, Stohr J (1980) Carbon K-edge fine structure in graphite foils and in thin-film contaminants on metal surfaces. Phys Rev B 21:2267–2273
80. Ching WY, Mo S-D, Chen Y, Si SC (2002) Calculation of XANES/ELNES spectra of all edges in Si3N4 and Si2N2O using a recently developed first-principles supercell method that includes the electron and core-hole interaction. J Am Ceram Soc 85(1):11–15
81. Marcelli A et al (2007) Local structure in SiO_2 glasses by oxygen K-edge XANES. Le J Phys Colloq 46(C8):C8-107–C8-112
82. Xiong G et al (2014) Different approaches to precise wavelength calibration of a flat-field grating spectrometer for laser-produced plasmas*. Phys Scr 89:65005
83. Mooney J, Kambhampati P (2014) Correction to get the basics right: Jacobian conversion of wavelength and energy scales for quantitative analysis of emission spectra. J Phys Chem Lett 5(20):3497–3497

Chapter 4
Time-Resolved XAS of Gas Dissociations

Time-resolved X-ray absorption spectroscopy has for long been experimentally limited to picosecond temporal resolution measurements in large-scale facilities. Recently, the expansion of high-harmonic generation table-top laser sources have led the femtosecond resolution of Ti:Sa oscillators to measurements in the extreme ultraviolet. This experimental chapter details the first results of the transient absorption setup developed in collaboration with the group of H.J. Wörner of ETHZ that are described in Chap. 2. This experiment measures the dissociation of molecules in the gas phase from the sulfur L-edges and the carbon K-edge in the soft X-rays. The temporal resolution of the setup allows to follow light-induced chemical reactions of the lowest electronic states with a few femtoseconds resolution. By comparing the calculated core-level transitions with the observed splitting of orbital states, the assignment of resolved structures, in accordance with the literature, allows to reconstruct the geometry changes of the molecules along their reaction path. As most of the experimental results and interpretations have already been published together with the group of H. J. Wörner [1], the descriptions and analysis given in this chapter provide a complementary view and some further discussions of the physical concepts involved in this experiment.

4.1 Introduction

The application of X-ray sources to the study of the structure of matter has led to some of the most prominent advances in science in the twentieth century. During the early twenty-first century, the temporal dimension has been added to X-ray measurements, both at synchrotrons [2] and through the recent development of free-electron lasers [3–6]. Aside of these fields of research, incoherent table-top hard X-ray sources have been applied to picosecond time-resolved studies [1, 7]. As an alternative approach,

C. Schmidt, *Time-Resolved Soft X-Ray Absorption Spectroscopy of Molecules in the Gas and Liquid Phases*, Springer Theses,
https://doi.org/10.1007/978-3-030-67838-8_4

high-harmonic generated (HHG) soft X-rays combine the advantages of full temporal and spatial coherence with perfect temporal synchronization.

The results presented in the following sections, represent the first successful femtosecond time-resolved experiment using soft-X-ray into the water window [8]. Using the experimental setup described in Chap. 3 with its 1.8 μm driver pulse configuration of high average power (2.5 W, i.e. 2.5 mJ at 1 kHz), a soft X-ray super-continuum ranging from 100 to 350 eV is generated by HHG in neon. Covering the chemically and biologically important K-edge of carbon, the laser source becomes a mean to study dissociation reactions of molecular cations in the gas phase. As a first attempt, molecules were targeted in the gas phase with strong-field interaction, representing the best option to generate a clear transient signal and demonstrate the feasibility of time-resolved X-ray absorption spectroscopy (TR-XAS) with table-top light sources. These choices were made first because the targeted molecules are directly regenerated between each measure in gas and secondly, because the signal induced by strong-field was expected to be easier to detect than for a resonant excitation. This development considerably extends pioneering works on transient absorption in the extreme ultraviolet [9–12], which were limited to photon energies below $\sim$100 eV, and therefore limited to the L-edge of silicon and the M- and N-edges of heavier elements. Recently, similar light sources reported measured molecular dynamics in the gas phase over the water window [13, 14].

Tetrafluoromethane (CF_4) has been the object of fundamental investigations in molecular chemistry for a long time because of its wide use in technological applications such as plasma dry etching of semiconductors, discharge switches, and gas-filled detectors [15, 16]. Besides, gaseous sulfur hexafluoride (SF_6) is an important electronic insulator used by industry as a substitute for PCB's, and large quantities are released into the atmosphere every year [17]. In the past decades, dissociation of CF_4 and SF_6 ions from the low-lying electronic states has been reported in many experimental methods, e.g., photoionization [18, 19], electron impact ionization [20–23] and velocity imaging photoionization coincidence [24–28]. If these previous techniques describe the valence-shell electronic configuration of CF_4 and SF_6, their different ionization channels and the respective dissociation fragments, none of them can provide high enough temporal resolution to extract qualitative information on the dynamics of the dissociation process. As a benchmark for the transient absorption system with sub-50 fs time resolution developed in this thesis, these two molecules were chosen for their high geometrical symmetry, known ionization energies (IE) and fast dissociation process.

Exploiting the sensitivity of X-ray absorption near-edge structure (XANES) spectroscopy to molecular conformation, the experiment follows the evolution of the unoccupied valence orbitals of molecules from the neutral to the final photodissociation reaction products after strong-field ionization.

4.2 Preliminary Density-Functional Theory Calculations

During the experiment preparation, density-functional theory (DFT) calculations of core–shell absorption spectra were computed for the targeted molecules and their fragments. For these preliminary calculations, the numerical StoBe (Stockholm-Berlin) program package was used to evaluate the molecular orbital electronic structure of the targeted transition. The algorithm is based on self-consistent solutions of the Kohn–Sham density functional theory equations using a linear combination of Gaussian type orbitals (LCAO) approach.

The DFT theory has become very popular in reason for its reduced computational resource requirement as compared to other ab-initio or semi-empirical theories based on the many-electron wave function. It has initially been developed to investigate the electronic ground-state structure of many-body systems in the condensed phases. Although the use of the DFT method is significantly increasing and was particularly attractive for these early calculations, its formalism introduces some limitations when describing charge-transfer systems due to the locality of the exchange and correlation (XC) functionals [29]. Moreover, a time-dependent (linear response) DFT formalism is generally required to accurately describe excited electronic state properties, where multiple electronic configurations are often involved. Indeed, if such properties could in theory be derived using an exact functional that maps the ground state density onto the desired property, in practice, only approximated functionals that favors an optimal arrangement of the electrons are used [29]. These considerations led to investigate, in a second time, more advanced quantum calculations based on the time-dependent DFT (TDDFT) theory. These calculations were realized by Yoann Pertot and Hans Jakob Wörner using the Amsterdam Density Functional (ADF) suite of programs at ETHZ.

The DFT simulations presented in this section were therefore mostly considered to evaluate the expected XANES shape modification and compared with the resolving power of the detection spectrometer.

Based on previous works on CF_4, [28, 30, 31] where the CF_4^+ ion was observed to dissociate into $CF_3^+ + F$ and other fragmentation channels, the expectation to observe an ionization-induced dissociation with the transient absorption setup was high. All the carbon-based ion fragments were therefore calculated with their corresponding spatial symmetry geometry basis (Fig. 4.1). The exchange and correlation effects were gathered using the generalized gradient approximation (GGA) exchange functional of Becke [32] and the GGA correlation functional of Perdew [33]. As the calculations generate oscillator strengths (linked to the cross sections, see Sect. 2.5.1) at specific energies and the experimental measurements record continuous spectra, the discrete result form of the Kohn–Sham equation is directly broadened with a Lorentzian function of 0.5 eV FWHM (Fig. 4.1, blue curve) [34, 35] instead of using an augmentation basis to expend the broadening representation over the edge.

The Lorentzian natural broadening of the electronic state typically accounts for the lifetime uncertainty of the excited state and the pressure effects [36]. To simulate the experimental spectral response of the G1 spectrometer configuration (see Sect. 3.5.5)

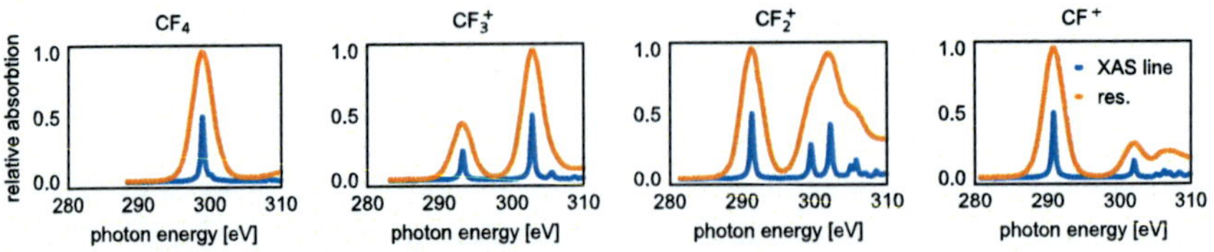

Fig. 4.1 Oscillator strength absorption of SXR photons at the carbon K-edge calculated for different dissociation fragments of CF_4. The blue curve correspond to 0.5 eV Lorentzian broadening and the orange curve is a Voigt broadening with $\gamma = 0.5$ eV and $\sigma = 3$ eV function parameters. The orange curve is vertically shifted for better visibility. The geometry basis used for each calculation are: T_d (CF_4), D_{3h} ($CF_3{}^+$), C_{2v} ($CF_2{}^+$), C_{2v} (CF^+)

with 3 eV resolution at the carbon K-edge, a Voigt function-based broadening was applied to the discrete oscillator strength values with 0.5 eV FWHM Lorentzian parameter, accounting for the natural broadening, and 3 eV FWHM Gaussian parameter (Fig. 4.1, orange curve), accounting for the spectrometer. The Voigt function is defined by:

$$V(E;\sigma,\gamma) = \int_{-\infty}^{\infty} G\big(E';\sigma\big)L\big(E-E';\gamma\big)dE' \tag{4.1}$$

where $G(E;\sigma)$ and $L(E;\gamma)$ are the Gaussian and Lorentzian functions and E is the energy. The pre-edge absorption band located at around 292 eV in $CF_3{}^+$, $CF_2{}^+$ and CF^+ fragments was therefore supposed to be distinguishable from the neutral CF_4 molecule.

As a similar dissociation process was previously observed with the SF_6 molecule. The corresponding calculation results are given in Fig. 4.2 and Fig. 4.3 for the sulfur L_{23} and L_1 edges respectively.

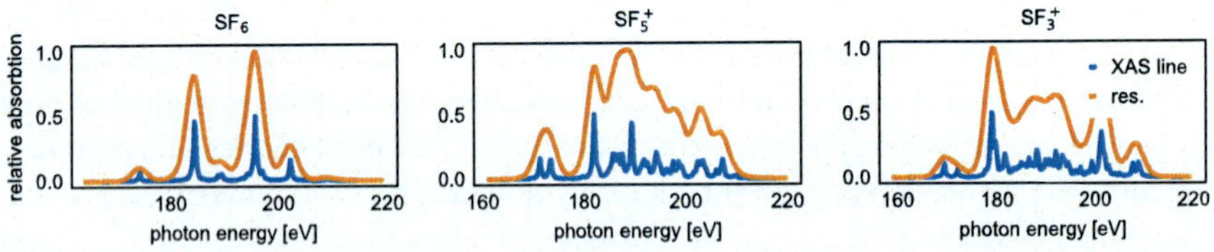

Fig. 4.2 Oscillator strength absorption of SXR photons at the sulfur L_{23}-edge calculated for different dissociation fragments of SF_6. The blue curve corresponds to 0.5 eV Lorentzian broadening and the orange curve is a Voigt broadening with $\gamma = 0.5$ eV and $\sigma = 3$ eV parameters. The orange curve is vertically shifted for better visibility. The geometry basis used for each calculation are: O_h (SF_6), D_{3h} ($SF_5{}^+$), C_{3v} ($SF_3{}^+$)

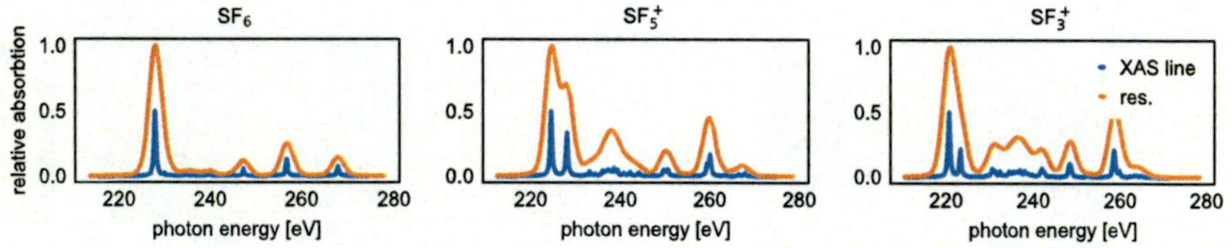

Fig. 4.3 Oscillator strength absorption of SXR photons at the sulfur LI-edge calculated for different dissociation fragments of SF6. The calculation and representation parameters are similar to Fig. 4.2

In the case of SF_6 fragments, the structure appears to be much more complicated than for the CF_4 with the appearance of multiple features between the main absorption peak in $SF_5{}^+$ and $SF_3{}^+$. Also, the first band calculated in SF_6 at about 172 eV splits in two bands in $SF_5{}^+$ and $SF_3{}^+$ that should not be directly resolvable with the spectrometer. However, dissociation of the molecule appears to be observable from the overall sulfur L_{23}-edge structure modification. Calculations of the L_1-edge spectral cross-section (Fig. 4.3) also shows a split of the main contributing peak that should be large enough to be directly measured experimentally.

4.3 Time-Resolved XANES of the CF_4 Carbon K-edge During Strong-Field Ionization Induced Dissociation

The experimental data were measured in Geneva using the setup introduced in Chap. 3 in its first configuration. Strong-field ionization of neutral CF_4 in the gas phase is induced by a 50 fs NIR pulse centered at 800 nm delivering $4 \cdot 10^{14}\,\text{W/cm}^2$ at focus position. The acquired data are normalized following the procedure detailed in Sect. 3.6. Two typical time scans are represented in Fig. 4.4, while the averaged XANES profile of the neutral (no pump pulse) CF_4 and post-ionized (300 fs after strong field) fragments are shown in Fig. 4.5 together with some spectral fitting attempts.

When looking at the time evolution of the absorbance in Fig. 4.4, the following observations are made. At $\Delta t = 0$ delay, corresponding to the maximum temporal overlap between the pump and the probe pulse, a redshift of about 0.1 eV and a small increase of absorbance from the main structure is observed. This spontaneous transient effect can be interpreted as a change in the XANES cross-section of $CF_4{}^+$ as compared to the neutral CF_4 during strong-field ionization or as a signature of the optical Stark effect [37], more precisely of the variable electromagnetic field (AC) Stark shift. Such effect can be observed [38, 39] when an intense laser radiation interacts with an atom and induces an energy shift $\Delta E = -\alpha(\omega_L) I_L/4$ of the ground state (where the polarizability $\alpha(\omega_L)$ depends on the dipole matrix elements and the laser frequency ω_L, and I_L is the laser intensity) [40]. At about 130 fs positive delays, a stronger increase of the absorbance and a larger red shift of its maximum are observed in the TR-XAS spectra shown in Fig. 4.4. After reaching a maximal absorbance, the

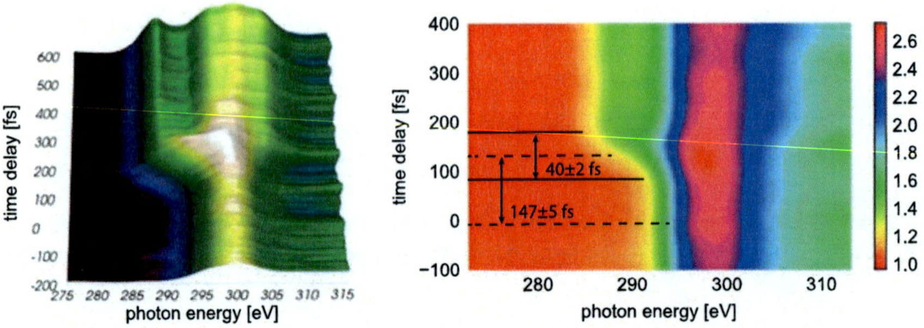

Fig. 4.4 Time-resolved transient absorption spectroscopy of CF_4 molecules in the gas phase. The color scale refers to the absorbance $A(\Delta t)$ as a function of the time delay Δt between the 800 nm pump and the SXR probe pulses. Negative time delays correspond to the SXR probe preceding the NIR pump. Left: 160 time-steps of 5 fs each from −200 fs to +600 fs. Every step is acquired over 20 s. Right: Similar experimental conditions with 80 time-steps after delay $\Delta t = 0$. Every step is acquired over 40 s

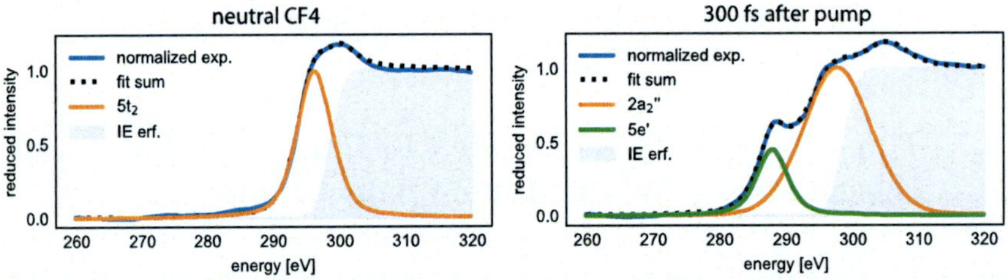

Fig. 4.5 XANES experimental absorption cross-section of CF_4 (left) and 300 fs after (right) strong field ionization. An error function is centered at 301.8 eV and represent the strong absorption edge induced by the ionization energy. Left: the measured spectrum is correctly fitted with a single Voigt function attributed to the C 1s to $5t_2$ transition. Right: the measured spectrum, of high $CF_3{}^+$ dominance, require two Voigt functions for good fitting. The pre-edge component is assigned to the C 1s to $5a_2''$ transition and the main peak to the C 1s to $5e'$ transition in $CF_3{}^+$

large and smooth main absorption structure progressively splits, and a strong pre-edge peak arises at 288 eV. At 200 fs delay, the pre-edge peak is stabilized, and the main absorption structure broadens on the blue side, showing some oscillations on the post-edge side. The same 10 eV band shift and main structure broadening have been observed in multiple scans.

For the data analysis, a $\beta = 0.5$ fraction of the neutral CF_4 absorption spectrum has been subtracted from the data set to account for partial ionization of the probed sample (see Sect. 3.6). The fraction of ionized molecules has been estimated using the ADK formula, for a hydrogen-like 1s orbital in a 50 fs, 800 nm pulse with a peak intensity of $4 \cdot 10^{14}\,\mathrm{W/cm^2}$ and an IE of 16.3 eV, corresponding to the vertical ionization energy of CF_4. This calculation predicts an ionization fraction of 90%. Although, considering that the p-orbitals of the fluorine atoms are dominating the HOMO, HOMO-1 and HOMO-2 of the CF_4 molecule (see Sect. 2.1.3), their ionization rate is expected to be significantly reduced as compared to the hydrogen 1s-orbital. Combined with the uncertainty on the pump and probe volume overlap, such estimation supports the determined ionization fraction of 50% ($\beta = 0.5$).

4.3.1 Details on Time Resolution

In order to quantify the time scales involved in the reaction process, it is important to distinguish the pump-probe pulse cross-correlation time from the intrinsic time evolution of the measured dynamics.

The experimental time resolution (cross-correlation) has been determined over multiple 0.3 eV wide energy intervals over the 284–289 eV range, corresponding to the appearance of the pre-edge absorption band assigned to CF_3^+. A Heaviside step function convoluted with a Gaussian function has been used to fit each of the extracted signals. Since narrow energy bands have been extracted for this analysis, it is assumed that the time broadening is dominated by the setup time resolution. All the individual Gaussian FWHM retrieved durations were averaged together in order to obtain a time resolution of 40 ± 5 fs.

The determination of the dissociation time is achieved by a similar fitting of the variation in energy as a function of time delay of the pre-edge feature. The center of mass of the pre-edge was used to extract the peak energy value at each time step (see Fig. 4.9a). The retrieved shift duration is 40 ± 2 fs and represents the experimental time appearance of the CF_3^+ pre-edge absorption band. As this time value is limited by the time resolution of the measurement, it only gives an upper limit of the time duration for the appearance of the CF_3^+ cation.

4.3.2 K-edge XANES Peak Assignment of CF_4 and CF_3^+

To date, multiple attempts to assign the carbon K-edge XANES dominant features of CF_4 have been proposed [30, 41]. According to its tetrahedral (T_d symmetry) geometry, the two lowest unoccupied orbitals (LUMOs) in the ground state are $5a_1$ and $5t_2$. These two states are formed from an admixture of the C sp^3 hybridized orbitals and F 2p orbitals and consist of four degenerate orbitals with antibonding character (σ*). As these orbitals appear to have very close binding energies [41], the assignment of the main feature preliminary calculated by DFT for CF_4 and shown in Fig. 4.1 require a deeper theoretical analysis of the C 1s core electron to valence transition in order to correctly describe the spectral profile. From the advanced TDDFT results of the neutral CF_4 it appeared that the transition from the C 1s to the $5a_1$ orbital is forbidden and that possible transitions towards a higher valence orbital ($6t_2$) can occur. The broad measured structure might therefore correspond to transitions to two overlapping states and was attributed to the $5t_2$ and $6t_2$ orbitals [8].

Table 4.1 CF_4 and CF_3^+ C K-edge XANES peak assignment together with the fitted Voigt parameter E_0 and FWHM and the corresponding preliminary DFT calculated values

Molecule (symmetry)	Assigned MO	E_0 (exp.) eV	E (DFT) eV	HWHM eV
CF_4 (T_d)	$5t_2$	296.3 (0.3)	298.2	6.2 (1)
CF_3^+ (D_{3h})	$2a_2''$	287.90 (0.03)	293.3	5.40 (0.08)
	$5e'$	297.70 (0.07)	303.1	11.9 (0.1)

In order to compare the results from calculations with the corresponding measured species, the averaged carbon K-edge XANES are normalized and fitted for CF_4 and CF_3^+. The spectrum of CF_4 was recorded without the pump pulse and, the spectrum of CF_3^+ was taken 300 fs after the pump pulse. The acquisitions were realized over 500 s in both cases and the results are shown in Fig. 4.5 together with an assignment of the main features.

In accordance with the DFT calculations presented in Sect. 4.2, the CF_4 spectrum is fitted with a single Voigt function and the CF_3^+ with two Voigt functions. To account for the strong absorption edge induced by the C 1s ionization energy previously calculated [42] and measured [43] at 301.8 eV, an error function centered at IE = 301.8 eV with 3 eV width parameter is used in both spectral fits [44]. The Voigt function $V(E; \sigma, \gamma)$ parameters are optimized using the Levenberg–Marquardt non-linear least-squares minimization method of the IFEFFIT's (standing for Interactive f_eff Fitting, where f_eff is the effective scattering amplitude coefficient when calculating X-ray absorption fine structure [45]) package [46]. The initial guesses for the line-width accounting for the natural broadening and the 3 eV resolution of the spectrometer at 300 eV (see Sect. 3.5.5) are $\gamma = 0.5$ and $\sigma = 1.3$. The retrieved E_0 and FWHM parameters are given in Table 4.1.

For these CF_4 and CF_3^+ "static" profiles, the summation of an error function with one or two Voigt functions was enough to explain the measured XANES. The retrieved FWHM of the different peaks (6.2, 5.4 and 11.9 eV respectively) are in remarkably good agreement with the literature [41, 47, 48] that describe broad features before the ionization energy. Moreover, if the absolute energy values of the extracted peaks are globally shifted as compared to the preliminary DFT calculations, the energy separation between the $2a_2''$ and the $5e'$ absorption bands of CF_3^+ of 9.8 eV was well predicted.

The observed changes in the absorption spectrum are the signature of symmetry lowering that occurs when the initially tetrahedral (T_d) CF_4 molecule dissociates into the trigonal planar (D_{3h}) CF_3^+ molecule. Indeed, arguments from group theory combined with dipole-selection rules show that the transition from the carbon 1s core orbital to a valence t_2 orbital in CF_4 must split into two transitions of type 1s to a_2'' and 1s to e' in CF_3^+.

Further insight accounting for Rydberg orbitals and partial valence to Rydberg mixing of the $5e'$ state of CF_3^+ was obtained by comparison with advanced ab-initio TDDFT calculations and are discussed in the following sections.

4.3.3 CF_4 Dissociation Paths

The CF_4 molecule has a tetrahedral (T_d symmetry) geometry. Its ground state electronic configuration is given in Fig. 2.1 from Sect. 2.1.3. The highest occupied molecular orbitals are $1t_1$, $4t_2$ and $1e$ of respectively 16.29, 17.51 and 18.54 eV vertical ionization energies [49]. Ionizing one electron from one of these valence orbitals leads to CF_4^+ in the X^2T_1, A^2T_2 or B^2E ionic state [28, 48]. As previously observed, ionization of CF_4 to each of these three lowest-lying CF_4^+ ionic states results in spontaneous dissociation into CF_3^+ and F [28, 31, 50]. Previous time of flight experiments of the ionic fragments released from electron impact ionization of CF_4 measured the different components, CF_3^+, CF_2^+, CF^+, F^+ and C^+ and reported the dominance of the CF_3^+ fragment [48].

In order to verify the dominating $CF_4^+ \rightarrow CF_3^+ + F$ dissociation under similar strong-field ionization conditions as measured by the reported TR-XAS experiment, the NIR pump pulse parameters and focusing geometry have been reproduced in a velocity-map-imaging spectrometer by Vit Svoboda and Yoann Pertot in the group of H.J. Wörner at ETHZ. The ion time-of-flight measurements shown in Fig. 4.6 are consistent with exclusive single ionization of the parent ion and a strong predominance of CF_3^+. The unstable CF_4^+ parent ion is not observed in these results because of its very short lifetime [51].

The typical time scale for ionization induced dissociation spans from the sub-femtosecond to a few picoseconds depending on the dissociation mechanism involved along the reaction path. From the experimental data of Fig. 4.4, the time between strong-field and full dissociation is measured to be less than 200 fs. By assuming a spontaneous ionization of CF_4 to CF_4^+ at $\Delta t = 0$ fs time delay, the complete

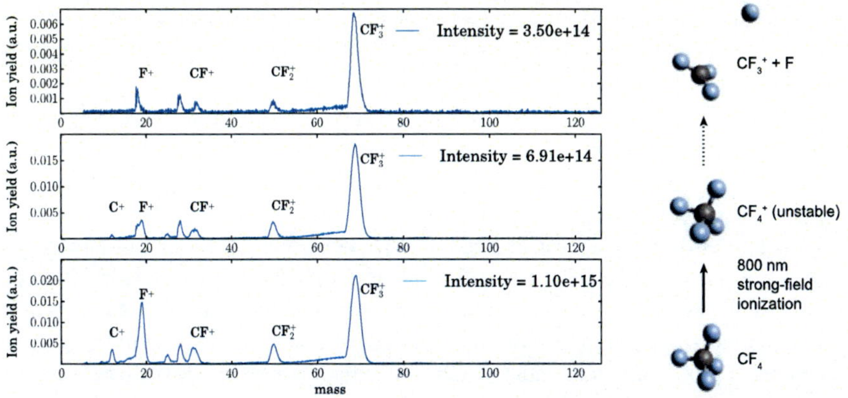

Fig. 4.6 Experimental measure of the CF_4 fragmentation with a velocity-map-imaging spectrometer for different NIR ionizing pulse intensities with similar parameters as used in the TR-XAS measurement together with a schematic representation of the molecule dissociation (right)

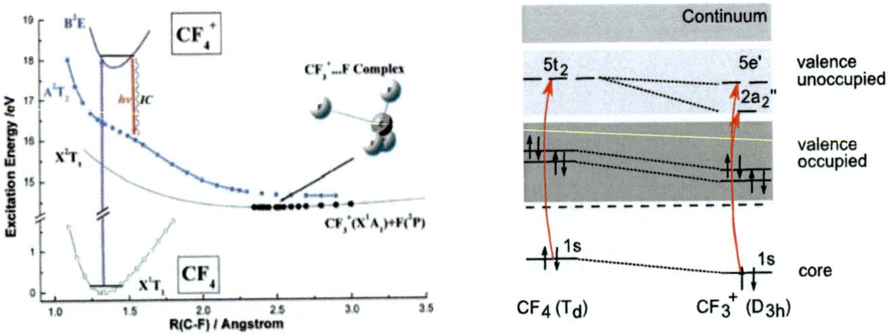

Fig. 4.7 Left: schematic potential energy curves of the minimum-energy reaction path from the three low-lying electronic states of CF_4^+ as a function of a single C-F internuclear separation. Reprinted from [28], with the permission of AIP Publishing. Right: core to valence XAS probed electronical transitions for CF_4 and CF_3^+

dissociation time given by the appearance of the CF_3^+ XANES signature accounts for 147 ± 5 fs.

Depending on the ionization pathway, different dissociation mechanisms have been recently discussed by [28] and are shown in Fig. 4.7. The first CF_4^+ ionic state (X^2T_1) potential energy surface was proposed to have a shallow minimum of C_{3V} symmetry (CF_3 + F complex) far away from the Franck–Condon region. This first ionization pathway is expected to have a long enough lifetime so that the CF_4^+ (X^2T_1) ions survive over a few vibrational periods before dissociation [52]. The investigation of such vibrational states is discussed in Sect. 4.3.5. The second ionic state A^2T_2 dissociation is expected to be very fast in agreement with the steeply repulsive potential curve in the Franck–Condon region that drives CF_4^+ to dissociate by Coulomb explosion. Finally, the dissociation through the third ionic state is suggested to follow a two-step dissociation mechanism, where the B^2E bound state first converts to the lower A^2T_2 state via internal conversion, then dissociates along its steep repulsive potential energy surface [28].

To confirm these previous results from the literature, the minimum-energy reaction path for the dissociation of CF_4^+ was calculated at ETH Zürich using the Gaussian09 package at the coupled-cluster singles-doubles CCSD/6-31G$*$ level of theory. The C-F internuclear separation has been fixed to a range of values between 1.32 and 5.3 Å while all other degrees of freedom have been relaxed in order to minimize the potential energy of the CF_3^+ + F electronic ground state along the dissociation path. The calculated energy curve precisely corresponds to the solid line of Fig. 4.7 (left) and exhibits the same minimum at 2.5 Å corresponding to a C_{3v} symmetry. As this minimum was found to be much smaller than the internal energy of CF_4^+ following strong-field ionization of CF_4 in its electronic ground state (i.e. the Franck–Condon region), all CF_4^+ molecules created by strong-field ionization were calculated to dissociate [8].

Although, a more advanced theoretical treatment would require the calculation of the three lowest ionic state potential-energy surfaces with full dimensionality

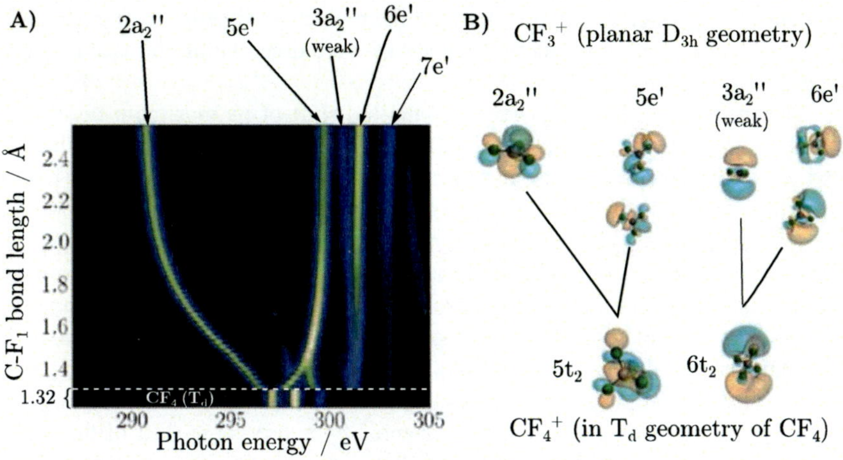

Fig. 4.8 **a** Calculated X-ray absorption spectra as a function of one C-F internuclear separation along the minimum-energy reaction path. The TDDFT method was used together with the LB94 functional and the QZ4P basis set. A linear intensity scale has been used. **b** Unoccupied orbitals characteristic after the X-ray transition, corresponding to the dominant absorption bands of panel (**a**). From [8]. Reprinted with permission from AAAS

(9 dimensions in the case of CF_4^+) and wave-packet propagation on these coupled surfaces, the dissociation through this minimum-energy reaction path corresponding to the first CF_4^+ ionic state (X^2T_1) has been simulated (see Fig. 4.8) and appears to be in very good agreement with the measured data.

With these observations, it is difficult to conclude on any favorable ionization pathway for the dissociation. Nevertheless, the sub-40 fs required for the CF_3^+ pre-edge appearance together with the "dead-time" observed after the pump pulse maxima ($\Delta t = 0$ delay) suggests a two-step dissociation mechanism with a very fast final reaction. Such interpretation can promote the dissociation pathways using the first ionic state CF_4^+ (X^2T_1) and its shallow minimum [8] or the third ionic state B^2E that converts to the lower A^2T_2 state via internal conversion and further dissociates to CF_3^+.

4.3.4 CF_4^+ (X^2T_1) Ionic State Dissociation Mechanism

The dissociation mechanism of the first ionic state X^2T_1 and its shallow minimum is elucidated by following the evolution in time of the main identified structures from the time-resolved measured absorbance and compare them with the unrestricted TDDFT calculations of the XANES for each geometry found along the minimum-energy reaction path. For these calculations, a Slater-type basis set of quadruple-zeta quality augmented with 4 polarization functions has been used for all the atoms of CF_4. As the wavelength of the SXR radiation of the probe pulse lies between

3.5 and 6 nm, which is much larger than the molecular dimensions (i.e. the F-C-F overall length of 0.3 nm), these calculations were done within the electric-dipole approximation.

The increase of the absorbance and the red shift of its maximum observed in the measured spectra at about 130 fs after the strong field ionization is attributed to the splitting of the $5t_2$ orbital into three nondegenerate orbitals in CF_4^+. This first splitting is a result of the Jahn–Teller effect stating that: "A nonlinear polyatomic system in a spatially degenerate electronic state distorts spontaneously in such a way that the degeneracy is lifted and a new equilibrium structure of lower symmetry is attained" [53]. The Jahn–Teller geometrical distortions therefore spontaneously remove the threefold degeneracy of the unequally occupied $5t_2$ orbital by reducing the symmetry of the molecule from T_d to C_{2v} at short ($\leq$1.6 Å) C-F bond lengths [31]. As the C-F bond length increases along the reaction path, the minimum-energy geometry changes to D_{3h} and is accompanied by a very large shift of the lowest-energy orbital assigned to the unoccupied $2a_2''$ orbital in CF_3^+ and corresponding to the strong pre-edge peak appearing at 288 eV in the measured spectra of Fig. 4.4. This large energy split mostly occurring toward the low energy side originates from the reduced energy of the unoccupied $2a_2''$ orbital of CF_3^+ as compared to the $5t_2$ orbital of CF_4. In parallel, the second- and third-lowest energy orbitals merge toward a single band, attributed to the $5e'$ orbital in CF_3^+, which explains the main absorption structure broadening on the blue side observed in the scan after 200 fs. Additionally, the appearance of two new absorption bands at 302 and 303 eV also contributes to this blue side broadening. Such structures, initially absent from the preliminary DFT calculations of Sect. 4.2, required these unrestricted TDDFT calculations to be revealed in reason of their initial Rydberg character. Indeed, when the fluorine atom detaches from the central carbon atom, the effective potential barrier created by the presence of the surrounding electronegative fluorine atoms becomes more transparent and the Rydberg orbitals with initially small overlap with the carbon 1s core orbital exhibit an increase of their absorption by acquiring partial valence character. By group symmetry, these bands are attributed to the CF_3^+ $6e'$ and $7e'$ orbitals of valence-Rydberg mixed character [8].

The presence of the shallow minimum in the minimum-energy reaction path suggests a residual binding interaction between the detaching fluorine atom and the CF_3^+ structure at equilibrium. To confirm that the planar D_{3h} symmetry is obtained in the dissociation process and with the intention of correctly describe the XANES at the end of the measured reaction (>300 fs), a restricted calculation on the closed-shell molecule CF_3^+ has been realized and compared to the open shell $CF_3^+ + F$ complex with very long C-F internuclear separations. The two calculations converged into similar absorption spectra, such that all the orbital symmetries of the final product were easily assigned by comparison. These comparative results also pointed out the importance in the choice of the exchange–correlation functional (LB94 in this case) for describing the asymptotic behavior of the electron close to the ionization threshold and for ionic species.

4.3.5 Transient Vibrational Modes Analysis

In order to investigate the presence of vibrational modes expected during the CF_4^+ X^2T_2 ion dissociation channel [28], the extracted pre-edge peak, shown in Fig. 4.9a, was used to find the energy center of mass of the peak at each time step. After the subtraction of an error function allowing removing the energy band shift contribution to the signal, the relative energy shift values of the absorption line assigned to the $2a_2''$ orbital of CF_3^+ were used to perform a sliding-window Fourier transform as a function of the delay. A 150 fs FWHM Gaussian time window has been used to extract the main frequencies as a function of time. The resulting spectrogram is shown in Fig. 4.9d, together with vibrational frequencies of CF_3^+ taken from ab-initio calculations [54] and spectroscopic measurements on isolated molecules in the gas phase [55].

The two main vibrational features extracted from the different CF_4 dissociation scans and visible in Fig. 4.9c, d are observed between 100 and 400 fs after the strong-field interaction. This indicates that some CF_3^+ vibrational modes are excited during the dissociation process. The first retrieved frequencies in the range of 750–850 cm^{-1} correspond to the previously calculated and measured 809 cm^{-1} frequency of the CF_3^+ umbrella mode with a_2'' symmetry [55]. The second feature visible between 1000 and 1200 cm^{-1} may correspond to the CF_3^+ a_1' breathing mode previously calculated at 1056 cm^{-1} [54].

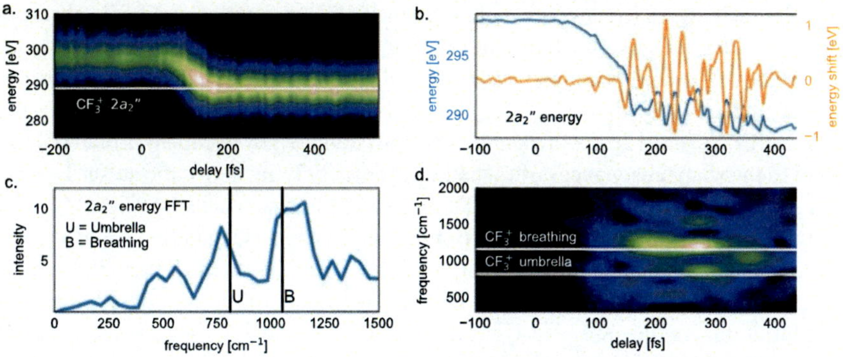

Fig. 4.9 Transient vibrational modes analysis. **a** Extracted $2a_2''$ absorption band from the experimental time scan of CF_4 dissociation. **b** Retrieved energy center of mass of the extracted band as a function of the time delay Δt. The blue curve is the energy where the orange curve is the relative energy shift after subtraction of an error function accounting for the band shift. **c** Fast Fourier transform of the $2a_2''$ energy shift signal. Some literature values of vibrational frequencies for CF_3^+ in its electronic ground state are indicated by the vertical lines. The umbrella mode line accounts for the previously measured 809 cm^{-1} frequency [55] and the breathing mode line corresponds to the previously calculated 1056 cm^{-1} frequency [54]. **d** Spectrogram of the $2a_2''$ energy shift signal with the same literature values. A 100 fs FWHM moving Gaussian window has been applied to the data in order to obtain the representation

The agreement between the experimentally observed Fourier components and the values found in the literature suggest that both, the umbrella and breathing modes of planar CF_3^+ undergo a few oscillations after the dissociation process. As the oscillation period of the breathing mode is 31.6 fs, about 10 oscillations are observed. The umbrella mode seems to oscillate over a shorter period and is less defined. Moreover, as the 40 ± 5 fs measurement time resolution is of similar time scale than the 41.2 fs period of the umbrella mode, the energy oscillatory signal of Fig. 4.9b might contain false features and more data with a higher time resolution (e.g. shorter pump pulse) are required to better distinguish between physical processes and measurement artifacts.

4.4 Time-Resolved XANES of the SF_6 Sulfur $L_{2,3}$-edge During Strong-Field Ionization Induced Dissociation

After the successful observations made in CF_4, the strong-field induced photodissociation of SF_6 was studied by looking at the sulfur L-edges in a similar way as for the CF_4 molecule. The strong-field ionization (Keldysh parameter $\gamma_K = 0.5$) of neutral SF_6 in the gas phase is induced by a 50 fs NIR pulse centered at 800 nm delivering $5 \cdot 10^{14}\ \mathrm{W/cm^2}$ at the focus position. Comparably as for the previous case, all the three lowest ionic states of SF_6^+ with respective vertical ionization energies of 15.7, 17.0 and 17.0 eV [56] are known to dissociate to $SF_5^+ + F$ [57].

For the data analysis process, a fraction (55%) of the measured SF_6 absorption spectrum has been subtracted from the whole data sets to account for the partial ionization of the probed sample (see Sect. 3.6). The measured XANES spectrum of the neutral SF_6, shown in Fig. 4.10 (left) before the $\Delta t = 0$ fs delay, is dominated by the sulfur $2p_{1/2}$ and $2p_{3/2}$ core shell transitions to the a_{1g} unoccupied orbital (LUMO) and to the shape resonances of t_{2g} and e_g symmetry lying above the sulfur $L_{2,3}$-edge as shown in Fig. 4.10 (right) [58]. These two shape resonances (see Sect. 4.4.2) are confined by the strong potential barrier formed by the 6 surrounding electronegative fluorine atoms [59]. In the case of SF_6, the presence of these transitions in the experimental XANES spectrum is explained by their strong overlap with the central sulfur atom. As they are both lying above the $2p_{1/2}$ and $2p_{3/2}$ ionization thresholds, they are contained within the cage (see Sect. 4.4.2).

After strong-field ionization, a global redshift is observed for all the spectral features. The a_{1g} band shifts by 2 eV, the t_{2g} band shifts by 2.5 eV and seems to split into two peaks separated by 6.5 eV, while the e_g structure also strongly shifts and broadens considerably. If the shift of all transitions appears to occur almost simultaneously, the splitting of the t_{2g} band is delayed by $\sim$30 fs. According to the advanced TDDFT calculations realized at ETH Zürich, this retarded pre-edge peak appearance is caused by a strong sensitivity of the t_{2g} shape resonance to bond angle changes. The rapid change of the bond angles occurring for S-F distances from 2.5 to 3 Å may cause a strong variation of the LUMOs energies close to the

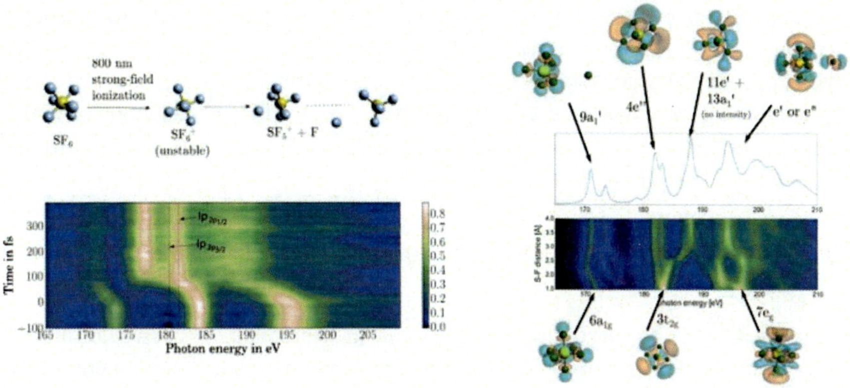

Fig. 4.10 Experimental TR-XAS scan of SF_6^+ together with its respective calculated absorption spectra along the dissociative reaction path. Left: Scheme of the dissociation process with a typical experimental scan. The color scale refers to the absorbance $A(\Delta t)$ as a function of the time delay between the 800 nm pump and the SXR probe pulses. Every 5 fs time-step is acquired over 40 s. Right: The calculated absorption spectrum of SF_5^+ is shown above the two-dimensional spectrogram of the complete dissociation. These advanced calculations were performed using the spin–orbit relativistic coupling TDDFT theory from the ADF suite of programs. The orbital shapes of the dominant absorption bands are shown with labels in order to illustrate the measured transformations

sulfur $L_{2,3}$-edge. Besides, the global observed red shift is dominated by a core-level shift of the sulfur atom. As a fluorine atom is moved away from the initial structure during the dissociation process, the 2p core orbitals shift to smaller (less negative) binding energies, lowering the photon energy of the relative absorbing transitions [8]. These insights are suggested by the advanced calculations performed using a spin–orbit relativistic coupling TDDFT theory with the ADF suite of programs. In this package, the exchange and correlation effects were gathered using the PW92 [60] LDA exchange functional and the GGA correlation functional of Perdew [33].

The broadening of the $7e_g$ shape resonance was numerically properly predicted (see Fig. 4.2) by the appearance of additional resonances on both sides, which partially merge to produce a broad absorption feature consisting of multiple shape resonances in SF_5^+. In the advanced TDDFT calculations, the split of the $3t_{2g}$ absorption band was also correctly reproduced and is visible in Fig. 4.10 (right). However, the two bands resulting from this splitting are experimentally observed to shift towards lower photon energies, whereas the calculation converged in one of the two bands shifting to higher energies. This inconsistency between experiment and theory is attributed to the limited accuracy of the advanced TDDFT calculation realized in this work and the complexity in describing strong-field induced dynamics.

The assignments of the observed XANES features during the time-resolved dissociation are done through symmetry group theory and by comparison with previous studies [58, 61, 62]. The SF_6^+ cation can dissociate into the SF_5^+ ground state from each of its three lowest electronic states [57]. SF_5^+, in its electronic ground state, has a trigonal bipyramidal geometry that belongs to the D_{3h} point group. In this symmetry, the transition from core to a_{1g} orbital of O_h group (SF_6) is attributed to a transition leading to an a_1' orbital. Analogously, the t_{2g} and e_g shape resonances of SF_6 correlate with a_1', e′ or e″ shape resonances in SF_5^+. These symmetry assignments were done by comparing the experimental observations to the advanced calculations shown in Fig. 4.10 (right).

4.4.1 Details on Time-Resolution

As in the case of CF_4, the time resolution of the measurement has been determined by integrating the signal at each delay over a narrow energy range, corresponding to the appearance or to a shift of an absorption band. The a_{1g} band intensity was integrated over the 168.5–169 eV energy range, the t_{2g} lower energy component appearance over the 175.9–176.5 eV energies, the t_{2g} main shift over 180.6–181.2 eV and the e_g energy shift between 187.6 and 188.2 eV. A Heaviside step function convoluted with a Gaussian function has been used to fit each of the extracted signals. The mean FWHM retrieved duration provides an experimental time resolution of 45 ± 7 fs. This time resolution is slightly worse than the value found when looking at the XANES of the carbon K-edge (CF_4, see Sect. 4.3.1). This difference can be induced by multiple factors. First, by considering the probe pulse duration. As the involved harmonic orders are lower at the sulfur $L_{2,3}$-edges (180 eV) than at the carbon K-edge (285 eV), more optical cycles of the NIR driving pulse contribute to the SXR generation. Nevertheless, with the high non-linearity of the HHG process, it is more suitable to believe that the time resolution of the experiment, typically given by the convolution of the pump and the probe pulses, is actually limited by the pump duration. During the strong-field interaction between the pump pulse and the gas sample, the instantaneous ionization rate of the sample strongly depends on its vertical ionization energy. Indeed, the ADK ionization rate calculation for the CF_4 molecule with IE = 16.3 eV shows that despite a 50 fs FWHM measured pump pulse duration, the FWHM of the envelope of the instantaneous ionization rate is only 24 fs. Doing a similar calculation for SF_6, gives a slightly larger temporal FWHM due to it respective lower vertical ionization potential of 15.7 eV. These considerations are therefore consistent with the slightly shorter time resolution found for the CF_4 carbon K-edge than for the SF_6 sulfur $L_{2,3}$-edges.

The determination of the dissociation time was achieved by a similar fitting of the retrieved peak energy shift values shown in Fig. 4.11 (left). The absorption peak energy positions have been extracted for each time step by fitting the signal intensity with Gaussian functions over restricted energy ranges corresponding to the considered structures. The positions of every Gaussian profile were then combined to reconstruct the evolution of the main measured XANES features over the dissociation process (see Fig. 4.11). If the normalized line shifts representation of Fig. 4.11 (right), shows no noticeable time delay between the dynamics of the individual observed shifts, the fitting analysis reveals time durations of 34 ± 4 fs for the a_{1g} band shift, 74 ± 10 fs for the t_{2g} lower energy component appearance, 40 ± 8 fs

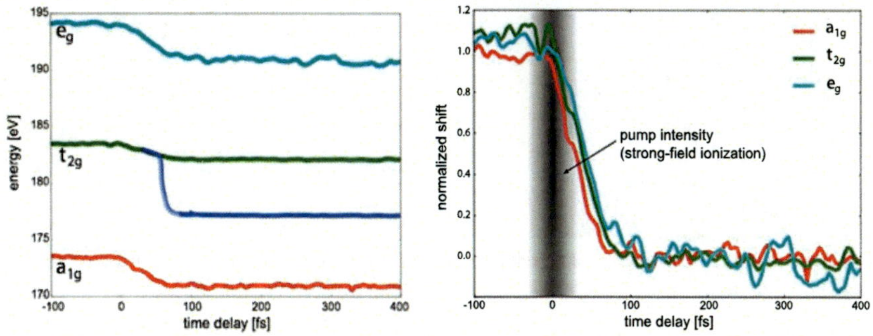

Fig. 4.11 Experimental retrieved central energy values of the absorption lines during the dissociation of SF_6^+ to $SF_5^+ + F$. The dotted blue line was artificially added and was not extracted from the measured data. Right: normalized time-dependent representation of the line shifts. The shadow black feature illustrates the excitation of the sample

for the t_{2g} main peak energy shift and 46 ± 4 fs for the e_g band energy shift. The retarded t_{2g} band splitting already discussed above and attributed to the different spatial structure associated to the one-electron orbitals of the t_{2g} shape resonance of SF_6 [8], thus provides a lower time limit of 74 ± 10 fs for the full dissociation in the conditions reported in this experiment.

4.4.2 Cage Effect and Shape Resonances

In chemistry, the "cage effect" usually describes how the properties of a molecule are affected by its surroundings in solvents [63]. It suggests that an individual particle must diffuse from its solvent cage in order to interact with other molecules. Over the years, this effect has been used to describe similar behaviors in atomic and molecular potential surface description. Typically, the effective potential barrier created by the presence of the surrounding electronegative fluorine atoms [59], forms a cage around the central carbon or sulfur atom, as shown in Fig. 4.12, of the CF_4 or SF_6 molecules, respectively. Therefore, a photoelectron excited from a sulfur or carbon core-level with energy higher than the ionization threshold can be temporally trapped in the cage before subsequent tunneling to the continuum.

When measuring XAS, a resonance describes the allowed promotion of a core electron into an unoccupied valence orbital. Such transition can occur towards an inner- or an outer-well state of the molecular potential. In molecules, such configuration typically arises when the electrons of the outer atoms produce a potential barrier along the bond directions as previously described for the cage effect of CF_4 and SF_6. Therefore, orbital states with an initial spatial distribution inside the potential barrier (inside the cage) are recognized as inner-well orbitals. Allowed transitions to these states are usually resulting in intense absorption features in XANES spectra [61]. However, orbital states located outer the cage lead to Rydberg state transitions that

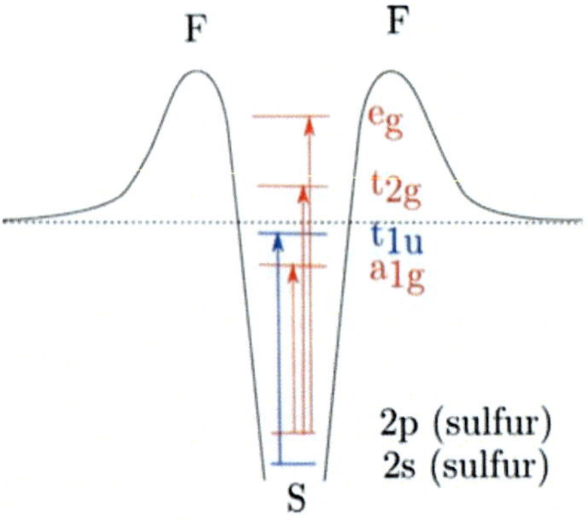

Fig. 4.12 Illustration of the cage effect in the case of SF_6. The t_{2g} and e_g shape resonances of energies above the ionization potential are trapped by the effective potential barrier created by the surrounding electronegative fluorine atoms

are generally weak in XANES signals because of the poor spatial overlap between the initial and final state orbitals [41]. When the core electron promotion leads to a quasi-bound state, inside the effective potential barrier, with energy above the absorption edge, as in the case of $3t_{2g}$ and $7e_g$ in SF_6, it is called a shape resonance [61].

The t_{2g} and e_g shape resonances are the dominating features above the L_{23}-edges. During the dissociation process, and with the detachment of a fluorine atom, the molecular cage becomes more transparent, so that the resonance becomes broader and more asymmetric in shape [58], as observed in the measured scan shown in Fig. 4.10 (left).

In the advanced TDDFT calculations realized for SF_6, the global observed red shift induced by a core-level shift of the sulfur atom when the cage opens, is under-estimated. Indeed, in order to describe such evolution of the core levels accurately, larger basis sets including diffuse basis functions are required, which lead to excessive calculation times.

4.4.3 Comparison of Typical Measured XANES with Synchrotron Data

To evaluate and characterize the quality and accuracy of these XANES measurements, it was natural to compare with previous experiments from mature synchrotron light sources. Typically, it allows for the energy calibration of the spectrometer and helps for the assignment and recognition of the observed features.

As an example, a typical static spectrum acquired over 60 s at the sulfur $L_{2,3}$-edges of SF_6, in the same conditions as reported for the time-resolved experiment, was compared to a XAS spectrum measured using SXR from the SX700/II monochromator of the Freie Universität Berlin (BESSY) [64]. The comparison between the two experimental spectra is shown in Fig. 4.13 (right). With normal emittance at BESSY, the typical resolution of the first order of diffraction is 45 meV FWHM at

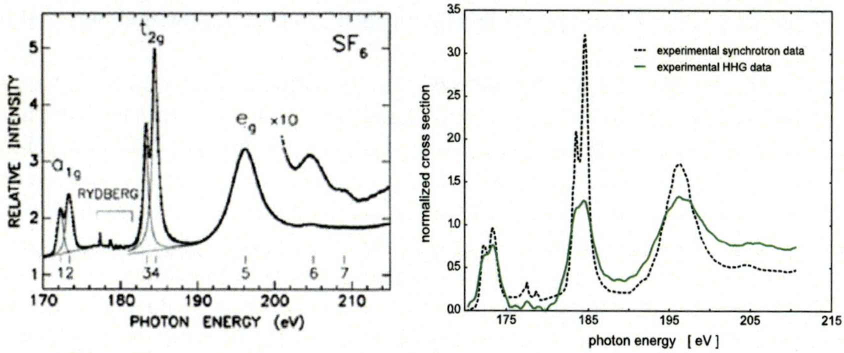

Fig. 4.13 Left: experimental photoabsorption spectrum of SF_6 near the sulfur $L_{2,3}$-edges in the gas phase measured with the synchrotron radiation from the SX700/II monochromator at BESSY. Reprinted figure with permission from [61], Copyright 2020 by the American Physical Society. Right: same data (dashed line) as the left figure compared to a static measurement of SF_6 with the same conditions as for the TR-XAS experiment reported in Sect. 4.4 (green line). The data are normalized on the a_{1g} band intensity

180 eV [61], where the experimentally determined resolution of the setup presented in this thesis is 600 meV at 180 eV.

When looking at the a_{1g} and t_{2g} features, it is apparent that the resolution achieved at the time of this experiment prevented the double structures observed at Bessy [61] to be resolved correctly. However, and despite the much lower spectral resolution of the tabletop setup presented in this thesis as compared to a synchrotron resolution, all the structures measured and shown in Fig. 4.13 (left) are observable in Fig. 4.13 (right, green curve). Moreover, the overall shape of the spectrum is very similar and the weak Rydberg states lying between the a_{1g} and the t_{2g} are resolved. This is a consequence of the natural linewidth calculated to be around 0.54 eV for a sulfur L_2 or L_3 orbital to valence transition, considering the excited state lifetime uncertainty only [36]. Such comparison demonstrates the ability to investigate the XANES signature of molecular species containing a core edge within the water window at the scale of a conventional optical laboratory.

4.5 Conclusion

By measuring the CF_4 and SF_6 fast photodissociation process with core to valence transition at a 40 fs time resolution, multiple temporal values and characterization of the ionic states were proposed. The comparison with initial density-functional theory calculations revealed the requirement of large basis sets including diffuse basis functions to correctly describe the evolution of the measured systems over their symmetry lowering. Generally, the good agreement with the advanced TDDFT simulations and the previously reported observations from synchrotron light sources demonstrates the sensitivity of the experimental setup to information provided by XANES spectroscopy.

The complete dissociation process of CF_4 has been observed from the carbon 1s core electronic level, and a maximal time duration of 152 fs has been determined until full molecular rearrangement. Moreover, the measured two-step like dissociation mechanism suggests multiple ionization channels to be involved before the fast and final geometrical transformation. By extracting the dominants Fourier components following the strong-field interaction with the laser pulse and temporally resolved over the reaction path, some previously reported vibrational modes were observed. Nevertheless, more qualitative insights on the spatial evolution would require a higher experimental time resolution or an analysis of the EXAFS for different geometries along the reaction path.

The strong shape resonances of SF_6 resulting from the cage effect induced by the fluorine atoms surrounding the sulfur central atom were identified and a large redshift of the XANES absorption lines was attributed to a core-level shift of the sulfur atom occurring when the cage opens up.

These results demonstrate the feasibility of TR-XAS with tabletop light sources and its potential in measuring the dynamics of electrons and nuclei in chemical reactions. Specifically, this method nicely complements other key techniques in molecular reaction dynamics, such as time-resolved photoelectron spectroscopy [65] and time-resolved high-harmonic spectroscopy [66]. TR-XAS can therefore be predicted to become a decisive technique for the investigation of molecular dynamics, such as those occurring at conical intersections [67, 68].

Owing to its sensitivity to elements, TR-XAS will enable time-resolved studies of electronic dynamics with atomic spatial sensitivity. Slightly increasing the signal-to-noise ratio of such sources would bring time-resolved extended X-ray absorption fine structure (TR-EXAFS) in combination with X-ray absorption near edge structure (XANES) experiments within reach, providing full structural and electronic information.

References

1. Ráksi F, Wilson KR, Jiang Z, Ikhlef A, Côté CY, Kieffer J (1996) Ultrafast X-ray absorption probing of a chemical reaction. J Chem Phys 104(15):6066–6069
2. Bressler C et al (2009) Femtosecond XANES study of the light-induced spin crossover dynamics in an iron(II) complex. Science 323(5913):489–492
3. Ackermann W et al (2007) Operation of a free-electron laser from the extreme ultraviolet to the water window. Nat Photonics 1(6):336–342
4. Emma P et al (2010) First lasing and operation of an ångstrom-wavelength free-electron laser. Nat Photonics 4(9):641–647
5. Ishikawa T et al (2012) A compact X-ray free-electron laser emitting in the sub-ångström region. Nat Photonics 6(8):540–544
6. Allaria E et al (2012) Highly coherent and stable pulses from the FERMI seeded free-electron laser in the extreme ultraviolet. Nat Photonics 6(10):699–704

7. Bargheer M et al (2004) Coherent atomic motions in a nanostructure studied by femtosecond X-ray diffraction. Science 306(5702):1771–1773
8. Pertot Y et al (2017) Time-resolved X-ray absorption spectroscopy with a water window high-harmonic source. Science (80-) 355(6322):264–267
9. Goulielmakis E et al (2010) Real-time observation of valence electron motion. Nature 466(7307):739–743
10. Schultze M et al (2014) Attosecond band-gap dynamics in silicon. Science (80-) 346(6215):1348–1352
11. Loh Z-H, Khalil M, Correa RE, Santra R, Buth C, Leone SR (2007) Quantum state-resolved probing of strong-field-ionized xenon atoms using femtosecond high-order harmonic transient absorption spectroscopy. Phys Rev Lett 98(14):143601
12. Attar AR, Bhattacherjee A, Leone SR (2015) Direct observation of the transition-state region in the photodissociation of CH3I by femtosecond extreme ultraviolet transient absorption spectroscopy. J Phys Chem Lett 6(24):5072–5077
13. Attar AR et al (2017) Femtosecond X-ray spectroscopy of an electrocyclic ring-opening reaction. Science (80-) 356(6333):54–59
14. Leitner T et al (2018) Time-resolved electron spectroscopy for chemical analysis of photodissociation: photoelectron spectra of Fe(CO) 5 , Fe(CO) 4 , and Fe(CO) 3. J Chem Phys 149(4):044307
15. Christophorou LG (1984) Electron-molecule interactions and their applications, vol 2
16. Pitchford LC, McKoy BV, Chutjian A, Trajmar S (1987) Swarm studies and inelastic electron-molecule collisions. In: Proceedings of the meeting of the fourth international swarm seminar and the inelastic electron-molecule collisions symposium, July 19–23, 1985, Tahoe City, California, USA
17. Dervos CT, Vassiliou P (2000) Sulfur hexafluoride (SF_6): global environmental effects and toxic byproduct formation. J Air Waste Manag Assoc 50(1):137–141
18. Walter TA, Lifshitz C, Chupka WA, Berkowitz J (1969) Mass-spectrometric study of the photoionization of C_2F_4 and CF_4. J Chem Phys 51(8):3531–3536
19. Creasey JC et al (1990) Nonradiative decay pathways of electronic states of group IV tetrafluoro and tetrachloro molecular ions studied with synchrotron radiation. J Chem Phys 93(5):3295–3306
20. Stephan K, Deutsch H, Märk TD (1985) Absolute partial and total electron impact ionization cross sections for CF 4 from threshold up to 180 eV. J Chem Phys 83(11):5712–5720
21. Deutsch H, Leiter K, Märk TD (1985) Unimolecular decay of CF_4^+ and CCl_4^+. Int J Mass Spectrom Ion Process 67(2):191–197
22. Kime YJ, Driscoll DC, Dowben PA (1987) The stability of the carbon tetrahalide ions. J Chem Soc Faraday Trans 83(2):403
23. Zhang W, Cooper G, Ibuki T, Brion CE (1989) Excitation and ionization of freon molecules. I. Absolute oscillator strengths for the photoabsorption (12–740 eV) and the ionic photofragmentation (15–80 ev) of CF_4. Chem Phys 137(1–3):391–405
24. Kinugawa T, Hikosaka Y, Hodgekins AM, Eland JHD (2002) New results on the dissociative photoionization of CF_4 and CCl_4. J Mass Spectrom 37(8):854–857
25. Biehl H, Boyle KJ, Smith DM, Tuckett RP (1997) The use of threshold photoelectron - fluorescence photon coincidence spectroscopy for the measurement of the radiative lifetimes of emitting states of CF_3X^+ (X = F, H, Cl, Br) ions. Chem Phys 214(2–3):357–366
26. Chim RYL et al (2002) Determination of the first dissociative ionisation energy of polyatomic molecules by threshold photoelectron photoion coincidence (TPEPICO) spectroscopy: application to CF4, SF6, SeF6, TeF6 and SF5 CF3. Surf Rev Lett 09(01):129–135
27. Garcia GA, Soldi-Lose H, Nahon L (2009) A versatile electron-ion coincidence spectrometer for photoelectron momentum imaging and threshold spectroscopy on mass selected ions using synchrotron radiation. Rev Sci Instrum 80(2):023102
28. Tang X et al (2013) Dissociation limit and dissociation dynamic of CF_4^+: application of threshold photoelectron-photoion coincidence velocity imaging. J Chem Phys 138(9):094306

29. Gross EKU, Dobson JF, Petersilka M (1996) Density functional theory of time-dependent phenomena. In: Density functional theory II. Springer, Berlin/Heidelberg, pp 81–172
30. Ueda K et al (1996) C 1s and F 1s photoabsorption and subsequent electronic decay of CH_4, CH_3F, CH_2F_2, CHF_3, and CF_4. J Electron Spectros Relat Phenomena 79:441–444
31. de la Vega JMG, San Fabián E (1991) Jahn-Teller effect and dissociation from the ground state of CF_4^+. Chem Phys 151(3):335–342
32. Becke AD (1988) Density-functional exchange-energy approximation with correct asymptotic behavior. Phys Rev A 38(6):3098–3100
33. Perdew JP (1986) Density-functional approximation for the correlation energy of the inhomogeneous electron gas. Phys Rev B 33(12):8822–8824
34. Cavalleri M, Odelius M, Nordlund D, Nilsson A, Pettersson LGM (2005) Half or full core hole in density functional theory X-ray absorption spectrum calculations of water? Phys Chem Chem Phys 7(15):2854
35. Odelius M, Cavalleri M, Nilsson A, Pettersson LGM (2006) X-ray absorption spectrum of liquid water from molecular dynamics simulations: asymmetric model. Phys Rev B 73(2):024205
36. Krause MO, Oliver JH (1979) Natural widths of atomic k and L levels, Kα X-rays lines and several Auger lines. J Phys Chem Ref Data
37. Bonch-Bruevich AM, Kostin NN, Khodovoi VA, Khromov VV (1969) Changes in the atomic Absorption spectrum in the field of a light wave. I. JETP 29(1):82–85
38. Chini M, Zhao B, Wang H, Cheng Y, Hu SX, Chang Z (2012) Subcycle ac stark shift of helium excited states probed with isolated attosecond pulses. Phys Rev Lett 109(7):073601
39. Chini M et al (2013) Sub-cycle oscillations in virtual states brought to light. Sci Rep 3(1):1105
40. Delone NB, Krainov VP (1999) AC Stark shift of atomic energy levels. Uspekhi Fiz Nauk 169(7):753
41. Ying JF, Leung KT (1994) Absolute transition probability measurement of valence-shell electronic structure of CF_4 and CCl_4 : bethe surfaces, and dipole-dominated preionization-edge valence and Rydberg states. J Chem Phys 100(10):7120–7134
42. Thomas TD, Shaw RW (1974) Accurate core ionization potentials and photoelectron kinetic energies for light elements. J Electron Spectros Relat Phenomena 5(1):1081–1094
43. Brown FC, Bachrach RZ, Bianconi A (1978) Fine structure above the carbon K-edge in methane and in the fluoromethanes. Chem Phys Lett 54(3):425–429
44. Outka DA, Stöhr J (1988) Curve fitting analysis of near-edge core excitation spectra of free, adsorbed, and polymeric molecules. J Chem Phys 88(6):3539–3554
45. Mustre de Leon J, Rehr JJ, Zabinsky SI, Albers RC (1991) Ab initio curved-wave X-ray-absorption fine structure. Phys Rev B 44(9):4146–4156
46. Newville M (2001) EXAFS analysis using FEFF and FEFFIT. J Synchrotron Radiat 8(2):96–100
47. Tronc M, King GC, Read FH (1979) Carbon K-shell excitation in small molecules by high-resolution electron impact. J Phys B At Mol Phys 12(1):137
48. Hossen K, Ren X, Wang E, Kumar SVK, Dorn A (2018) An (e, 2e+ ion) study of electron-impact ionization and fragmentation of tetrafluoromethane at low energies. Eur Phys J D 72(3):43
49. Holland DMP et al (2005) An experimental and theoretical study of the valence shell photoelectron spectrum of tetrafluoromethane. Chem Phys 308(1–2):43–57
50. Creasey JC et al (1993) Fragmentation of valence electronic states of $CF^+{}_4$ and $SF^+{}_6$ studied by threshold photoelectron-photoion coincidence spectroscopy. Chem Phys 174(3):441–452
51. Fiegele T, Hanel G, Torres I, Lezius M, Märk TD (2000) Threshold electron impact ionization of carbon tetrafluoride, trifluoromethane, methane and propane. J Phys B At Mol Opt Phys 33(20):4263–4283
52. Das A, Miller JS, Poliakoff ED, Lucchese RR, Bozek JD (2007) Vibrationally resolved photoionization dynamics of CF4 in the D 2A1 state. J Chem Phys 127(4):044312
53. Jahn HA, Teller E (2013) Stability of polyatomic molecules in degenerate electronic states I. Orbital degeneracy. In: Applied group theory, vol 161, no 905, pp 233–255
54. Bowman JM, Huang X, Harding LB, Carter S (2006) The determination of molecular properties from MULTIMODE with an application to the calculation of Franck–Condon factors for photoionization of CF_3 to. Mol Phys 104(1):33–45

55. Dossmann Soldi-Lose H, Garcia GA, Nahon L, de Miranda BKC, Alcaraz C (2012) Comprehensive vacuum ultraviolet photoionization study of the CF3(●) trifluoromethyl radical using synchrotron radiation. J Chem Phys 136(20):204304
56. Holland DMP et al (1995) An experimental and theoretical study of the valence shell photoelectron spectrum of sulphur hexafluoride. Chem Phys 192(3):333–353
57. Creasey JC et al (1993) Fragmentation of valence electronic states of $CF^{+}{}_{4}$ and $SF^{+}{}_{6}$ studied by threshold photoelectron-photoion coincidence spectroscopy. Chem Phys 174(3):441–452
58. Flesch R et al (2013) Gas-to-cluster effects in S 2p-excited SF6. J Chem Phys 138(14):144302
59. Dehmer JL, Dill D, Wallace S (1979) Shape-resonance-enhanced nuclear-motion effects in molecular photoionization. Phys Rev Lett 43(14):1005–1008
60. Perdew JP, Wang Y (1992) Accurate and simple analytic representation of the electron-gas correlation energy. Phys Rev B
61. Hudson E et al (1993) High-resolution measurements of near-edge resonances in the core-level photoionization spectra of SF6. Phys Rev A At Mol Opt Phys 47(1):361–373
62. Ying JF, Daniels TA, Mathers CP, Zhu H, Leung KT (1993) Absolute transition probability measurement of nondipole valence-shell (7–70 eV) electronic transitions of SF6 by angle-resolved electron energy loss spectroscopy. J Chem Phys 99(5):3390–3399
63. Franck J, Rabinowitch E (1934) Some remarks about free radicals. Trans Faraday Soc 30(1930): 120–130
64. Domke M et al (1992) Performance of the high-resolution SX700/II monochromator. Rev Sci Instrum 63(1):80–89
65. Suzuki T (2006) Femtosecond time-resolved photoelectron imaging. Annu Rev Phys Chem 57:555–592
66. Wörner HJ, Bertrand JB, Kartashov DV, Corkum PB, Villeneuve DM (2010) Following a chemical reaction using high-harmonic interferometry. Nature 466(7306):604–607
67. Polli D et al (2010) Conical intersection dynamics of the primary photoisomerization event in vision. Nature 467(7314):440–443
68. Wörner HJ et al (2011) Conical intersection dynamics in NO_2 probed by homodyne high-harmonic spectroscopy. Science 334(6053):208–212

Chapter 5
Pulse Peak Power Optimization with Filamentation

This chapter presents the HHG driving pulse optimization results. The intensity and short pulse requirements to reach higher soft X-ray energies and better conversion ratio are discussed and different approaches are introduced. The previous, demanding, time-resolved X-ray absorption spectroscopy experiment in the gas phase leaned towards the option for an easy pulse shortening scheme. The adopted two-stage pulse broadening by filamentation coupled to a very simple post-compression technique is environmentally robust and leads to a 0.2 TW sub-two-cycle short-wavelength infrared pulse. The latter can drive high-order harmonics beyond the oxygen K absorption edge with similar flux conditions as obtained during the successful gas dissociation measurements at the carbon K-edge.

5.1 Introduction

After the demonstration of the soft X-ray (SXR) transient absorption (TA) measurements of gas dissociation presented in Chap. 4 [1], the wish to extend the high harmonic generation (HHG) cut-off over the oxygen K-edge and cover the full water window, required the replacement of the generation gas. As expressed by the cut-off energy law given by Eq. 2.18, combining long wavelength λ drivers with a high ionization potential IP gas can lead to drastic extension of the SXR spectrum. Nevertheless, replacing neon by helium as generation medium required to increase the laser peak intensity to reach an equivalent Keldysh parameter, and achieve sufficient flux for sample measurement TA experiments. For a given pulse energy, the temporal shortening of the electric field envelope leads to a maximization of the HHG flux as it results in higher peak intensity and reduces the number of optical cycles [2]. The arising intensity gradient thus allows electrons of the generation gas to survive in the ground state up to laser fields many times higher than the binding Coulomb field,

C. Schmidt, *Time-Resolved Soft X-Ray Absorption Spectroscopy of Molecules in the Gas and Liquid Phases*, Springer Theses,
https://doi.org/10.1007/978-3-030-67838-8_5

leading to higher ionization rates. This was made possible by the idler pulse optimization described in this chapter and the development of the high-pressure generation cell described in Sect. 3.4.1.

Producing few-cycle short-wavelength infrared (SWIR) pulses via filamentation of femtosecond laser pulses in gases has been described in numerous studies after the improvement of tabletop amplified laser sources capable of reaching the critical power for Kerr self-focusing in noble gases ($\sim 10^{10}$W) [3–6]. Some striking results reported highly efficient pulse compression in argon gas, down to less than 10 fs at 800 nm, with a conversion efficiency of 85% [4]. Alternatively, some research groups have developed gas-filled hollow fibers with pressure gradient to produce sub-10 fs pulses with post-compressor [7]. Following those achievements at 800 nm wavelength, similar pulse compression techniques at longer wavelengths were also developed [8–12].

The physics of filamentation received increased attention as it enables a cost-efficient approach and robustness for intense laser pulse compression. Femtosecond pulses of peak powers close to the self-focusing threshold P_{cr} (see Eq. 2.13) indeed produce a single filament in which the spectrum broadens and typically acquires positive group delay dispersion (GDD) from self-phase modulation (SPM). In some cases, self-compression processes have been observed [3–5, 13, 14], revealing the presence of nonlinear effects that counterbalance the positive induced chirp. As the critical power P_{cr} scales with λ^2 [6, 15], longer wavelength reduces beam collapsing together with plasma density and its role in counterbalancing Kerr self-focusing.

In a classical Ti:Sa pulse compression scheme, the positive chirp acquired by SPM is compensated with a suitable negative dispersion technique to obtain a short, transform limited pulse. Experimentally, this can be achieved with the aim of prisms or a reflective compressor. While operating in the SWIR ($\lambda > 1.4\ \mu$m) spectral range, common optical materials such as fused silica (FS) or CaF_2 offer an even simpler way to compensate for the residual positive chirp with their anomalous dispersion curve [8, 16]. The amount of negative introduced GDD being directly proportional to the length of the medium. In practice, during filamentation, ultrashort pulses are also subject to higher-order dispersion and compensating only for GDD can lead to pulse pedestals and sometimes pulse splitting, which strongly affect the final pulse quality [17]. Therefore, the use of anomalous dispersion from bulk materials allowing for a simultaneous compensation of GDD and third-order dispersion (TOD) [8, 16] has to be carefully considered in order to find the best combination between self-guided propagation nonlinear effects and post-compression.

5.2 Filamentation of the 1.8 μm OPA Idler in Argon

The spectrum of the 1.8 μm idler pulses is broadened by self-guiding propagation in a first 3.5 m long cell isolated by two 3 mm thick CaF_2 windows cut at Brewster's angle and filled up to 3 bar of argon. To reduce plasma formation and its associated losses and instabilities, a 2 m long focal length silver mirror loosely focuses the beam for the first stage of spectral broadening. The beam is then collimated using a similar 2 m focal length silver mirror and the residual chirp is compensated in a FS glass plate to temporally compress and reach maximum peak intensity before the second stage. The pulse is then launched into a second broadening stage consisting of an identical argon-filled cell. The gas pressure in each cell can be independently adjusted. The beam is finally collimated by a 3 m focal length silver mirror to slightly increase its diameter, and the spectrum is monitored using an InGaAs spectrometer.

The 45 fs pulses centered around 1.8 μm from the parametric amplifier reach a maximum peak power of 5×10^{10} W which is equal to the critical power P_{cr} for 1 bar of argon pressure at 1.8 μm wavelength, considering a nonlinear refractive index $n_2 = 1 \times 10^{-19}(p/p_{atm})$ cm^2/W [19] and a Gaussian beam. On account of the very structured spatial profile of the pulse at OPA output, and the white-light continuum generated while argon pressure approaches 1.3 bar in the first cell, the assumption that the peak power is actually lower than estimation for a perfect Gaussian profile can be done. At this regime, Kerr self-focusing is expected to be sufficiently weak so that the diffraction prevents pulse collapsing and its associated break-up in multiple filaments. The focal spot diameter under vacuum is estimated at 600 μm, leading to an upper peak intensity of 0.3 PW/cm^2. To quantify the energy losses due to the propagation in the two broadening stages, the mean power is measured at the output of each stage for different pressures of argon. The pressure is gradually increased from vacuum in both cells to 2.5 bar in the first cell. Then, the pressure in the second cell is increased up to 2 bar while keeping the first cell at 1 bar. The two-stage setup transmission is measured to be more than 85% while the ionization introduces virtually no losses as depicted in Fig. 5.1c, d. The losses are therefore mainly attributed to Fresnel coefficients of the cell windows, which confirms the low plasma density generated in the filaments.

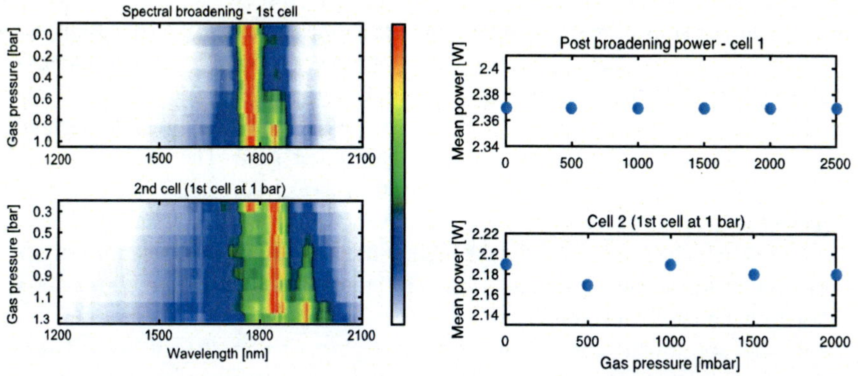

Fig. 5.1 Evolution of the pulse spectral broadening with increasing pressure in the first (**a**) and second gas cell (**b**) for the two-stage filament compression. For the second broadening stage (**b**), the pressure in the first cell is fixed at 1 bar. **c** Output pulse average power measurement at the output of each filament cell as a function of argon pressure in the **c** first gas cell and **d** second gas cell. Reprinted with permission from [18] © The Optical Society

The spectral broadening dependence on argon pressure is shown in Fig. 5.1a, b for the two filament stages. From 0.6 bar of gas in the first cell, a distinct second peak formation centered around 1850 nm appears in the spectrum while the broadening tends to slightly blue shift the overall spectrum center of mass. The main broadening process involved is via the Kerr nonlinearity n_2 induced SPM. A time-dependent nonlinear phase $\phi_{\mathrm{NL}} = 2\pi n_2 L E(t)^2/\lambda_0$ (L is the propagation distance, λ_0 is the carrier wavelength, n_2 is directly proportional to the gas pressure and $E(t)$ is the laser electric field) is accumulated while the pulse propagates, so that new frequencies ($\Delta\omega = -\frac{\mathrm{d}}{\mathrm{d}t}\phi_{\mathrm{NL}}$) are produced in the spectrum. The time-dependent intensity envelope thus produces red-shifted components near the leading edge of the pulse and blue shifted ones at the trailing edge. Pure SPM broadening exhibits oscillatory behavior, the number of peaks being associated with the maximal acquired nonlinear phase ϕ_{MAX} [20]. The origin of these peaks is the production of the same chirp within the pulse at different times and their interference. For a symmetric pulse envelope, a SPM broadened pulse should then exhibit symmetric peaks around the central carrier frequency. The additional effect of self-steepening, however, breaks this symmetry. As the refractive index $n(t) = n_0 + n_2 I(t)$ is larger in the center of the pulse than on the edges, the trailing part of the pulse propagates faster than the center, inducing a pulse steepening and eventually an optical shock [21]. This temporal asymmetry, in turn, modifies the spectrum as the steepness of the slopes on both edges of the pulse become asymmetric. Consequently, the intensity of the red-shifted peak is strongly enhanced as compared to the blue-shifted ones. On the other hand, the spectral broadening on the blue side is wider than on the red side [20]. This prominence of the red-shifted SPM peak at 1850 nm, as well as the broader blue wing, are well observed in Fig. 5.1, as the pressure increases from 0.6 bar to 1 bar. Recently, similar red-shifting behaviors have been shown with few-cycle 800 nm pulses in the filamentation regime [22], in the same spectral range by filamentation in krypton [23] and also with a hollow core fiber [9]. In particular, the self-steepening process and its associated red-shifting appears significant for further compression by a transparent bulk material [9]. Increasing the filling argon pressure above 1.5 bar deteriorates the pulse-to-pulse stability and the spatial profile. By coupling with the second propagation cell, final optimal pulse characteristics were obtained by limiting the broadening process of the first stage to the spectrum achieved at 1 bar. The broadening in the second filament stage follows similar description, as a clear redshift of the main peak together with blue shift broadening are observed as soon as the pressure is increased. At 1.3 bar, which corresponded to the highest pressure for stable operation, the spectrum covers the wavelength range from 1200 to more than 2100 nm with a 400 nm FWHM bandwidth and a main peak at 1960 nm.

Besides spectral broadening, the spatial interplay of Kerr self-focusing and plasma self-defocusing has been pointed to be responsible for filament self-compression [4, 5, 24]. Indeed, the accumulation of nonlinear effects occurring in filamentation leads to an important restructuring of the pulse time profile. Self-focusing, self-steepening, and SPM instantaneous effects tend to accumulate the laser energy to the ascending side of the pulse, while the photoionization effect tends to cut off the trailing part. Possibly, their combination can lead to significant time reduction of the pulse. With the loose self-guiding geometry adopted for this setup and the low-induced plasma density, such phenomenon is expected to be limited. The measured pulse duration at the output of the 1st and 2nd filament stages of respectively 59.2 fs and 57.5 fs together with the measured FROG traces (see Figs. 5.4 and 5.5) indicate that such self-compression, if existing, is limited to a small blue side region of the pulse and is not sufficient to efficiently compress all spectral components.

5.3 Filamentation of the 2.1 μm OPA Idler in Krypton

In order to take advantage of the $U_p \propto \lambda^2 \cdot I_S$ wavelength dependence and increase the flux at the high SXR spectrum energies, the setup was adapted to be used with the 2.1 μm OPA idler and the broadening cells were reduced to 2 m length to better fit on the experiment table. The pulses are focused with a 1.25 m concave mirror to an estimated 450 μm focal diameter under vacuum and is re-collimated by a 2 m focal length mirror after the second stage. The 45 fs FWHM measured pulses centered around 2.1 μm reach a lower peak power of 2.8×10^{10} W because of the reduced efficiency conversion of the OPA for longer wavelengths (see Sect. 3.2.1).

Considering a nonlinear refractive index $n_2 = 9 \times 10^{-20}(p/p_{atm})\,\text{cm}^2/\text{W}$ for argon [19] at 2.1 μm, the critical power rises to $P_{cr} \cong 7.3 \bullet 10^{10}$ W and is beyond the capacity of the OPA source. This deficit of peak power to initiate the filament can be compensated by increasing the argon pressure to 3 bar or by replacing the gas. Krypton, for instance, presents a higher nonlinear index coefficient $n_2 \cong 27 \times 10^{-20}(p/p_{atm})\,\text{cm}^2/\text{W}$ [25, 26]. The critical power is therefore reduced to 2.4×10^{10} W at 1 bar allowing for self-guided propagation with reduced pressure and thinner Brewster cell windows.

As shown in Fig. 5.2, comparable spectral broadening is reached with 1.2 bar of krypton and 3.2 bar of argon, confirming previous comparison with ultrashort Ti:Sa pulses [27]. The best broadening conditions are obtained by direct observation of the generated SXR spectrum. As in the case of the 1.8 μm pulse (Sect. 5.2), the optimal pressure parameters for pulse broadening before high harmonic generation are found to be near the threshold for white-light generation, occurring when the pulse peak power just exceeds the medium critical power P_{cr} for self-focusing [28].

5.4 Spatial Mode Cleaning

By looking at the far-field pattern of the beam after the two filamentation stages, the emerging SWIR pulse spatial profile shows an intense quasi-Gaussian core surrounded by a weaker annular SWIR radiation. Moreover, the imaged beam shows a more symmetric and smoother intensity profile after filamentation. The presence of the surrounding radiation ring pattern coupled to the blue-shifted SPM induced

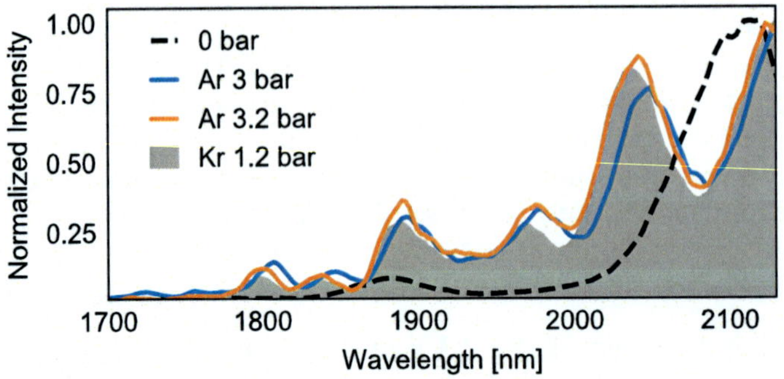

Fig. 5.2 2.1 μm pulse spectral broadening. The black dashed curve is the spectrum at OPA output, and the filled grey area is the spectrum after 2-stage broadening in 1.2 bar of krypton. The spectrum of the pulse displays an important SPM broadening accumulated over nonlinear propagation. The two argon curves are shown for comparison. Only the blue side of the spectrum is measurable with InGaAs spectrometer

wavelengths appearing during broadening may be a mark of the conical emission phenomenon [29, 30]. The surrounding radiation ring is visible in Fig. 5.3b which compares the spatial beam profile of the SWIR before and after filamentation.

This feature is characteristic of optimized filamentation schemes, where self-cleaning of the spatial mode occurs [31] and where radial symmetry is observed and discussed in many studies at similar propagation regime $P_{\text{laser}} \sim P_{\text{cr}}$ [32–34]. It appears that the mode cleaning effect is mainly due to the spatial dependence of the Kerr induced nonlinear refractive index $n_2 I(t)$ to the incoming pulse intensity distribution. Simulations also show that mode self-cleaning in gases can be associated with the generation of nonlinear X-waves [35]. These X-waves were initially described being nondiffracting and nondispersive extended solutions to the NLSE [36]. In filaments, high-intensity X-waves can result from the amplification of new phase-matched frequencies in three- and four-wave mixing processes, and are triggered by the conical emission [37]. Recent simulations demonstrated that such phenomenon

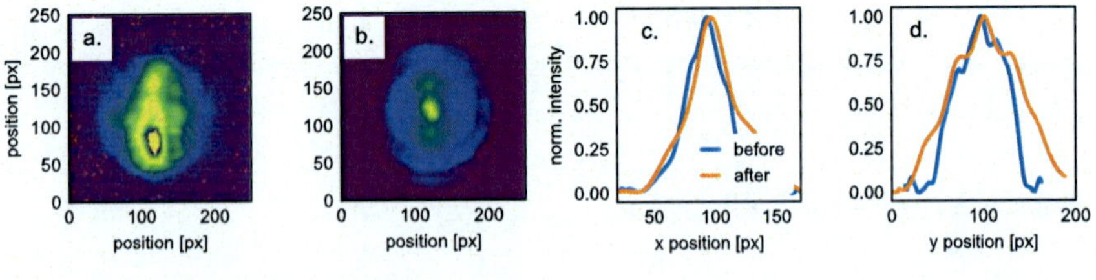

Fig. 5.3 1.8 μm spatial beam profile in the far-field. Measured at the output of the OPA (**a**) and after the two-stage filamentation setup (**b**). The x and y scales are given in pixels and the relative dimensions may be not conserved. The horizontal (**c**) and vertical (**d**) profiles are also represented for before (blue) and after (orange) filamentation

can happen even without ionization nor filamentation. The symmetrization is therefore only dependent on the Kerr effect that generates transverse spatial wave vectors, redistributing the energy and reducing the high spatial frequencies [31].

5.5 Sub-Two-Cycle Pulse Compression

The temporal characterization of the 1.8 μm pulse is performed using a dispersion-free third-harmonic generation optical gating (THG-FROG, see Sect. 3.3.2) for different pressures of argon and FS plate thickness. The pressure in the two broadening gas cells is optimized in order to achieve the broadest final spectrum after the two compression stages. This is done by moderately broadens the spectrum in the first stage in order to preserve clean temporal and spatial beam profiles. Figure 5.4 (Fig. 5.5) displays the measured FROG traces and associated autocorrelation (AC) signals for the best 1 bar (1.3 bar) pressure condition found in the 1st (2nd) filamentation cell (respectively).

After the first broadening stage, the about 300 nm FWHM broad spectrum supports a bandwidth-limited 14 fs pulse duration. Using the Eq. 2.5 for GVD chirp compensation applied to the 41 fs (59.2 fs AC) emerging pulse, the estimated amount of FS required to maximally compress the pulse is 2.8 mm as represented in Fig. 5.6 (left). The shortest measured 14.8 fs (20.9 fs AC) pulse for 2 mm of FS is therefore in good agreement with the calculated value. After compensation, the upper limit for the peak power is 1.5×10^{11} W, which corresponds to 3 times the critical power for self-focusing in argon at 1 bar pressure.

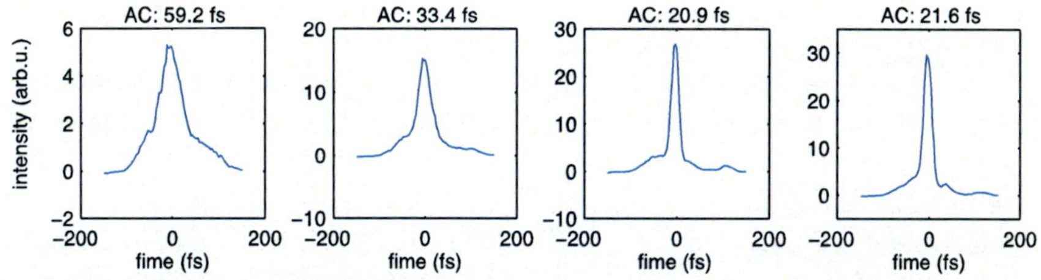

Fig. 5.4 Measured THG-FROG traces and AC pulse duration after the 1st filamentation cell and for 0, 1, 2 and 3 mm FS plate compression. For a Gaussian pulse, the intensity envelope FWHM duration is given by $\frac{\tau_{AC}}{\sqrt{2}}$ (where τ_{AC} is the AC FWHM)

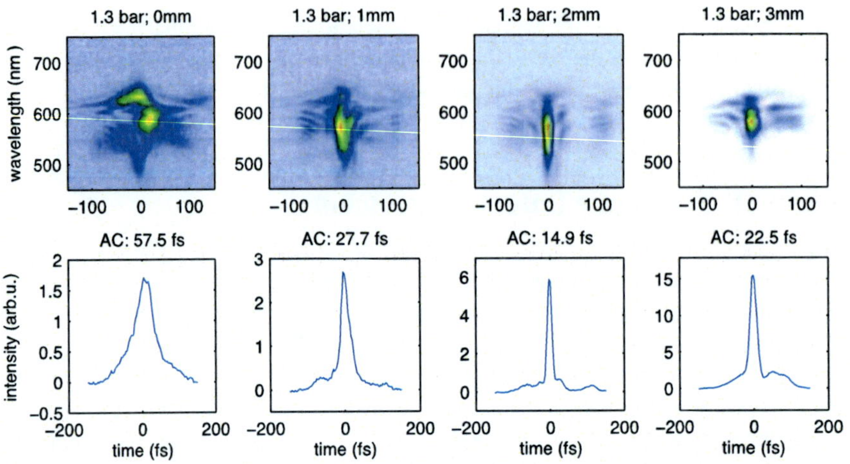

Fig. 5.5 Measured THG-FROG traces and AC pulse duration after 2nd filamentation cell and for 0, 1, 2 and 3 mm FS plate compression. The first cell is filled with 1 bar of argon and the pulse is compressed with 2 mm FS plate before the second stage leading to a 14.8 fs injected pulse

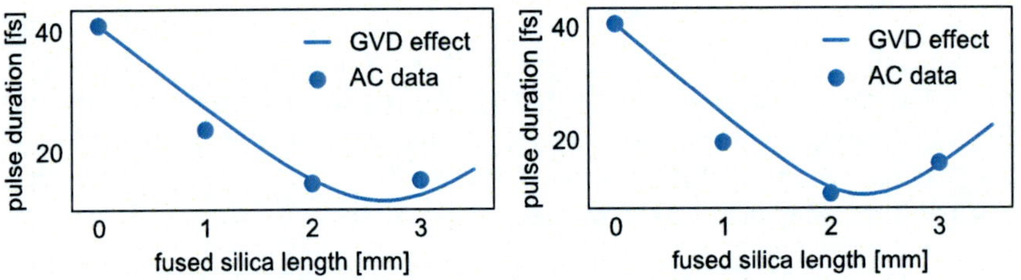

Fig. 5.6 GVD effect of 0, 1, 2 and 3 mm thick FS windows on the pulse duration. The solid curve is calculated from Eq. 2.5 and the data are measured by AC after 1st (left) and 2nd (right) filamentation stages. The AC signals and respective FROG traces are shown in Figs. 5.4 and 5.5

The subsequent filament stage broadens the spectrum to about 400 nm FWHM, supporting a bandwidth-limited 11 fs pulse duration. Again, the 2.2 mm estimated amount of FS required to maximize the compression of the released 40 fs pulse (see Fig. 5.6 (right)) correctly agree with experimental data.

As already introduced, the use of FS to take advantage of its anomalous dispersion and compensate for the positive chirp induced during filamentation also influence the TOD. Indeed, all glasses present positive TOD over their entire transmission range. For instance, the 2 mm length FS plate used to compress the 1.8 μm pulse after filamentation, introduces 600 fs^3 TOD which stretches a bandwidth-limited pulse duration from 10 fs to more than 12 fs [20]. Uncompensated TOD thus leading any femtosecond pulse to deviate from its transform-limited temporal profile even for zero GDD. Although the retrieved temporal profile shown in Fig. 5.7 reveals some pedestals, typically revealing residual TOD in the pulse, the main peak duration of

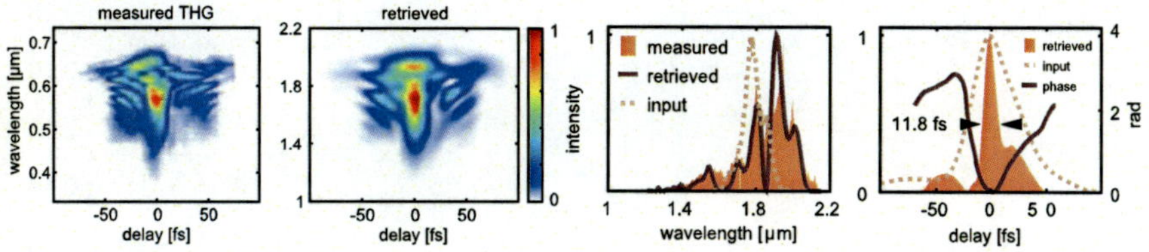

Fig. 5.7 THG-FROG reconstruction of the optimized compressed 1.8 μm pulse. Left: THG-FROG measured trace and the multi algorithm retrieved trace. Right: normalized comparison in the spectral and time domain of the OPA idler (input) and the same pulse after broadening and compression (measured). The input spectrum is centered at 1.79 μm with a 145 nm bandwidth and the measured spectrum shows a main peak shifted at 1.96 μm and is 400 nm wide at FWHM. The retrieved phase (right axis) is given for the compressed pulse only. Reprinted with permission from [18] © The Optical Society

11.8 fs suggests the presence of negative TOD after filamentation. The latter being attributed to the self-steepening effect, as previously discussed by Schmidt et al. [8, 38], and also observed for spectral broadening in hollow core fibers [9].

The final temporal characterization of the 1.8 μm pulse is carried out with full spatial coupling of the pulse with the THG-FROG. The retrieved pulse spectrum is in good agreement with an independent and direct measure with the InGaAs spectrometer. The retrieved temporal profile shows that the pulses are compressed by a factor of 4, down to 11.8 fs as summarized in Fig. 5.7. With these characteristics, the peak power of the pulse reaches as much as 0.2 TW, for an average power of 2.1 W.

5.6 Extended SXR Spectrum Generated in Helium

The combination of 0.2 TW peak power with short intensity rise time allows to efficiently generate harmonics in helium. The HHG geometry describes in Sect. 3.4.1 leads to a high enough intensity for strong tunnel-ionization process, whereas the reduced free-electron density in the high-pressure gas target before the arrival of the main electric field increases the high energy side of the converted signal levels over the detection limit [39–41]. The strong intensity gradient of the driver determining the maximum field strength that atoms are exposed to before the ejection of an outer electron and the dramatic drops of their polarizability.

The SXR generated spectrum shown in Fig. 5.8 is calibrated in energy and intensity following the procedure described in Sect. 3.5.4 and the absorption lines observed from TiO_2 nanoparticles deposited on a Si_3N_4 substrate of 50 nm thickness. As comparison, a typical HHG spectrum generated in neon with the uncompressed pulse is shown to emphasize the SXR photon energy extension. For an equivalent energy of 2 pJ per laser pulse around the carbon K-edge region, the helium-based harmonic continuum cut-off reaches 570 eV, representing a full coverage of the water window. The overall stability combined with the kHz pulse repetition rate

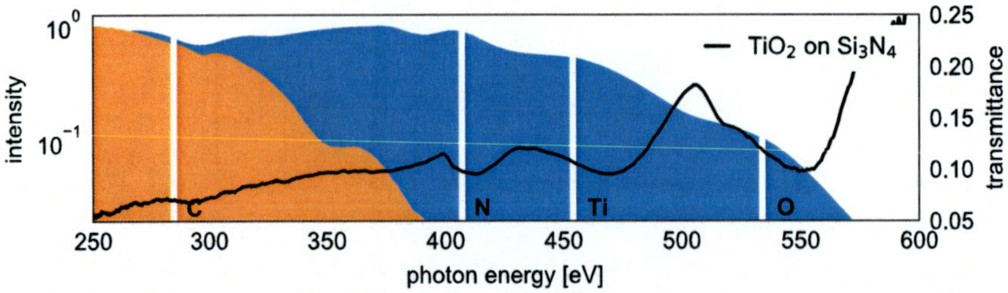

Fig. 5.8 SXR extended high harmonic spectrum after compression of the driver. In orange, the SXR generated in neon with the uncompressed 45 fs, 1.8 μm pulse. In blue, the spectrum generated in helium with the 11.8 fs, 1.8 μm short pulse. The generation is achieved with different target geometries and are overlapped to emphasis the 200 eV increase of the cut-off. The black curve corresponds to the transmittance of TiO_2 nanoparticles deposited on a Si_3N_4 substrate. The white lines indicate the energies of the carbon (C) K-edge at 285 eV, the nitrogen (N) K-edge at 407 eV, the titanium (Ti) L-edge at 454 eV and the oxygen (O) K-edge at 534 eV. Adapted from [18]

allows for transient absorption measurement in similar conditions and time scale as previously reported in Chap. 4 [1]. For instance, the static TiO_2 nanoparticles transmitted spectrum acquisition was averaged over 30 s.

When generating high harmonics with the 2.1 μm pulse, the cut-off extends to more than 600 eV as shown in Fig. 5.9. Though the SXR overall flux is globally lower than with the 1.8 μm wavelength because of the reduced efficiency conversion of the OPA source and the reduced single-atom yield scaling with $\lambda^{-5.5}$ [42], the high-energy region of the spectrum presents a higher signal-to-noise ratio so that the measurement of the oxygen K-edge (534 eV) is improved. This is depicted by the TiO_2 nanoparticle measurement realized with similar experimental conditions as with the 1.8 μm pulse (Fig. 5.8, solid line).

While HHG with long driver pulses typically results in the emission of a train of atto-bursts resulting from the interaction of the atomic gas with each wave cycle of the electric field, the optimized sub-two optical cycle SWIR pulse limits the recombination to an almost single wave cycle as shown in Fig. 2.5 (right) [44, 45]. However,

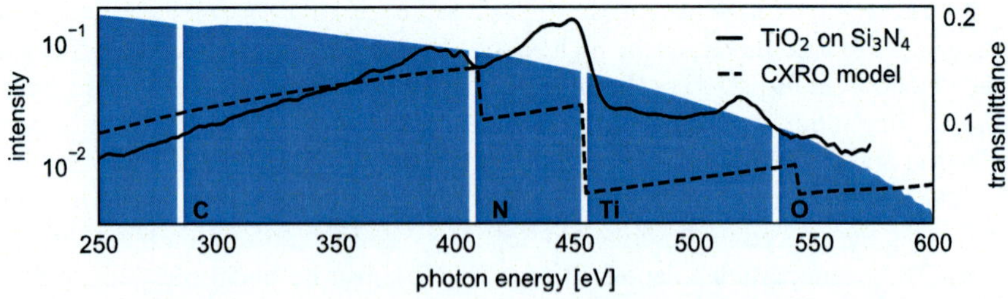

Fig. 5.9 SXR high harmonic spectrum generated in helium with the 2.1 μm short pulse. The generation is achieved in 4 bar of helium over 4 mm. The black curve corresponds to the transmittance of TiO_2 nanoparticles deposited on a Si_3N_4 substrate as for results presented by Fig. 5.8. The dashed curve is extracted from the CXRO database [43]

single SXR pulse extraction and relative attochirp characterization is outside the scope of this work and should be investigated in the future.

Finally, the conservation of the phase relation between the carrier wave and the pulse envelope over the filamentation process [3], validates the full compatibility of the pulse presented here with CEP stabilization. The f-2f interferometer feedback loop, presented in Sect. 3.3.3, that compensates for CEP drifts in the last amplification stage of the OPA can therefore be coupled to the output of the compression setup in order to stabilize the cut-off region of the SXR spectrum for increased pump-probe signal-to-noise ratio.

5.7 Discussion

Generation of high harmonic isolated pulses in the water window relies on high ionization potential gases such as helium and long wavelength sub-two-cycle driving sources. For time-resolved spectroscopic investigations like SXR transient absorption, a high flux is greatly appreciated, which requires high peak power SWIR sources with high repetition rates. In particular, recent investigations based on post-compression in hollow core fibers have shown that both high flux and cut-off frequencies at the oxygen K-edge [46, 47] are achievable.

While gas-filled hollow core fibers represent a key technique to produce ultrashort pulses, they still contain serious limiting factors. The maximal supported energy for damage threshold is about 1 mJ, while pointing instabilities of the input beam directly induce unwanted energy and pulse parameter fluctuations at output [48], requiring beam pointing stabilization for good coupling. As an attractive alternative, the filamentation-based pulse compression setup presented in this work and associated smooth HHG spectra covering the whole water window provides comparable performances. As compared to hollow core fibers, filamentation provides significant advantages while coupled to demanding experiment such as SXR transient-absorption spectroscopy. Due to the absence of waveguide in the broadening scheme, the alignment is greatly simplified, and the induced optical damage probability is significantly reduced. In addition, the good transmission efficiency of the full setup as compare to hollow core capillary-based schemes [8–11], allows to take advantage of most of the high average power delivered by the commercial laser system. The two-stage pulse compression arrangement [3, 49] allows achieving larger broadening and cleaner pulse shapes as the group delay dispersion is temporally optimized in the middle of the process. Moreover, the reduced required overall cell dimension allows for loose focusing geometries in a standard laboratory room.

The results presented in this chapter demonstrate that self-guided schemes are suitable for efficient pulse compression and generation of clean SWIR driving pulses for HHG covering the full water window. Actually, moving to a longer wavelength provides a factor 5 in the critical power (from 800 nm to 1.8 μm) before modulational instability amplifies, while operating in the SWIR ($\lambda > 1.4$ μm) spectral range allows simple compensation of the chirp by the anomalous dispersion of common optical materials. By analyzing the evanescent pulse in the far field, it is obvious that the self-mode-cleaning effect significantly improves the beam quality of the pulse, leading to better focusability and enhanced HHG. During the characterization of the pulse compression, the best conditions for a smooth spectrum and temporal profile were always found for gas pressures set in order to be at the threshold for white-light generation in the second stage of filamentation.

References

1. Pertot Y et al. (2017) Time-resolved x-ray absorption spectroscopy with a water window high-harmonic source. Science (80), 355(6322):264–267
2. Brabec T, Krausz F (2000) Intense few-cycle laser fields: frontiers of nonlinear optics. Rev Mod Phys 72(2):545–591
3. Hauri CP et al (2004) Generation of intense, carrier-envelope phase-locked few-cycle laser pulses through filamentation. Appl Phys B 79(6):673–677
4. Stibenz G, Zhavoronkov N, Steinmeyer G (2006) Self-compression of millijoule pulses to 7.8 fs duration in a white-light filament. In: 2006 conference on lasers and electro-optics and 2006 quantum electronics and laser science conference, pp 1–2
5. Skupin S et al (2006) Self-compression by femtosecond pulse filamentation: experiments versus numerical simulations. Phys Rev E Stat Nonlinear Soft Matter Phys
6. Couairon A, Mysyrowicz A (2007) Femtosecond filamentation in transparent media. Phys Rep 441(2):47–189
7. Suda A, Hatayama M, Nagasaka K, Midorikawa K (2003) Generation of 9-fs 5-mJ pulses using a hollow fiber with pressure gradient. In *OSA* trends in optics and photonics series
8. Schmidt BE et al (2010) Compression of 1.8 μm laser pulses to sub two optical cycles with bulk material. Appl Phys Lett 96(12):121109
9. Béjot P, Schmidt BE, Kasparian J, Wolf JP, Legaré F (2010) Mechanism of hollow-core-fiber infrared-supercontinuum compression with bulk material. Phys Rev A At Mol Opt Phys 81(6)
10. Silva F, Teichmann SM, Cousin SL, Hemmer M, Biegert J (2015) Spatiotemporal isolation of attosecond soft X-ray pulses in the water window. Nat Commun 6:6611
11. Cousin SL, Silva F, Teichmann S, Hemmer M, Buades B, Biegert J (2014) High-flux table-top soft x-ray source driven by sub-2-cycle, CEP stable, 1.85-μm 1-kHz pulses for carbon K-edge spectroscopy. Opt Lett 39(18):5383–5386
12. Vozzi C et al (2006) High-energy, few-optical-cycle pulses at 1.5 μm with passive carrier-envelope phase stabilization. Opt Express 14(21):10109
13. Kosareva OG et al (2008) Optimization of a femtosecond pulse self-compression region along a filament in air. Appl Phys B Lasers Opt
14. Bergé L, Skupin S, Steinmeyer G (2008) Temporal Self-restoration of compressed optical filaments. Phys Rev Lett 101(21):213901
15. Bergé L, Skupin S, Nuter R, Kasparian J, Wolf J-P (2007) Ultrashort filaments of light in weakly ionized, optically transparent media. Reports Prog Phys 70(10):1633
16. Demirdöven N, Khalil M, Golonzka O, Tokmakoff A (2002) Dispersion compensation with optical materials for compression of intense sub-100-fs mid-infrared pulses. Opt Lett 27(6):433
17. Demircan A, Kroh M, Bandelow U, Hüttl B, Weber HG (2006) Compression limit by third-order dispersion in the normal dispersion regime. IEEE Photonics Technol Lett
18. Schmidt C et al (2018) High-order harmonic source spanning up to the oxygen K-edge based on filamentation pulse compression. Opt Express 26(9):11834
19. Zahedpour S, Wahlstrand JK, Milchberg HM (2015) Measurement of the nonlinear refractive index of air constituents at mid-infrared wavelengths. Opt Lett 40(24):5794
20. Agrawal G (2001) Nonlinear fiber optics. Elsevier Science

21. Panagiotopoulos P, Whalen P, Kolesik M, Moloney JV (2015) Super high power mid-infrared femtosecond light bullet. Nat Photonics 9(8):543–548
22. Ahmad I, Bergé L, Major Z, Krausz F, Karsch S, Trushin SA (2011) Redshift of few-cycle infrared pulses in the filamentation regime. New J Phys 13(9):93005
23. Driever S et al (2013) Tunable 1.6–2 μm near infrared few-cycle pulse generation by filamentation. Appl Phys Lett 102(19):191119
24. Zaïr A et al (2007) Spatio-temporal characterization of few-cycle pulses obtained by filamentation. Opt Express
25. Brée C, Demircan A, Steinmeyer G (2010) Method for computing the nonlinear refractive index via Keldysh theory. IEEE J Quant Electron 46(4):433–437
26. Rice JE (1992) Frequency-dependent hyperpolarizabilities for argon, krypton, and neon: comparison with experiment. J Chem Phys 96(10):7580–7586
27. Nisoli M, De Silvestri S, Svelto O (1996) Generation of high energy 10 fs pulses by a new pulse compression technique. Appl Phys Lett 68(20):2793–2795
28. Corkum PB, Rolland C, Srinivasan-Rao T (1986) Supercontinuum generation in gases. Phys Rev Lett 57(18):2268–2271
29. Faccio D, Porras MA, Dubietis A, Bragheri F, Couairon A, Di Trapani P (2006) Conical emission, pulse splitting, and x-wave parametric amplification in nonlinear dynamics of ultrashort light pulses. Phys Rev Lett 96(19):193901
30. Béjot P, Kasparian J (2011) Conical emission from laser filaments and higher-order Kerr effect in air 36:4812–4814
31. Milián C et al (2015) Laser beam self-symmetrization in air in the multifilamentation regime. J Phys B At Mol Opt Phys 48(9):94013
32. Prade B et al (2006) Spatial mode cleaning by femtosecond filamentation in air. Opt Lett 31(17):2601–2603
33. Liu W, Chin SL (2007) Abnormal wavelength dependence of the self-cleaning phenomenon during femtosecond-laser-pulse filamentation. Phys Rev A 76(1):013826
34. Moll KD, Gaeta AL, Fibich G (2003) Self-similar optical wave collapse: observation of the townes profile. Phys Rev Lett 90(20):203902
35. Kolesik M, Wright EM, Moloney JV (2005) Interpretation of the spectrally resolved far field of femtosecond pulses propagating in bulk nonlinear dispersive media. Opt Express 13(26):10729
36. Conti C et al (2003) Nonlinear electromagnetic X waves. Phys Rev Lett 90(17):170406
37. Porras MA, Parola A (2009) X and Y waves in the spatiotemporal Kerr dynamics of a self-guided light beam. Opt Commun 282(4):644–652
38. Schmidt BE et al (2014) Intense few-cycle infrared laser pulses at the advanced laser light source. Chin J Phys 52(1):537–545
39. Chang Z, Rundquist A, Wang H, Murnane MM, Kapteyn HC (1997) Generation of coherent soft x rays at 2.7 nm using high harmonics. Phys Rev Lett
40. Spielmann C et al (1997) Generation of coherent X-rays in the water window using 5-Femtosecond laser pulses. Science (80), 278(5338) :661–664
41. Schnürer M et al (1998) Coherent 0.5-keV x-ray emission from helium driven by a sub-10-fs laser. Phys Rev Lett
42. Tate J, Auguste T, Muller HG, Salières P, Agostini P, DiMauro LF (2007) Scaling of wave-packet dynamics in an intense midinfrared field. Phys Rev Lett 98(1):13901
43. Henke BL, Gullikson EM, Davis JC (1993) X-ray interactions: photoabsorption, scattering, transmission, and reflection at E = 50–30,000 eV, Z = 1–92. At Data Nucl Data Tables 54(2):181–342
44. Sola IJ et al (2006) Controlling attosecond electron dynamics by phase-stabilized polarization gating. Nat Phys
45. Goulielmakis E et al (2008) Single-cycle nonlinear optics. Science (80)
46. Teichmann SM, Silva F, Cousin SL, Hemmer M, Biegert J (2016) 0.5-keV Soft X-ray attosecond continua. Nat Commun 7:11493 EP
47. Johnson AS et al (2018) High-flux soft x-ray harmonic generation from ionization-shaped few-cycle laser pulses. Sci Adv 4(5):eaar3761

48. Kornelis W et al (2004) Single-shot dynamics of pulses from a gas-filled hollow fiber. Appl. Phys. B 79(8):1033–1039
49. Hauri CP, Guandalini A, Eckle P, Kornelis W, Biegert J, Keller U (2005) Generation of intense few-cycle laser pulses through filamentation—parameter dependence. Opt Express

Chapter 6
Experimental Results of TR-XAS in Liquids

This chapter presents TR-XAS measurement results in the liquid phase. This chapter presents TR-XAS measurement results in the liquid phase. By coupling a vacuum flat jet interaction chamber with the HHG water window source described in Chap. 2 and taking advantage of the cut-off extension presented in Chap. 4, the induced reaction is tracked simultaneously from the carbon and nitrogen K-edges at sub-100 fs time resolution. Measurements were performed on the liquid pyridine (C_5H_5N) excited by ultrashort NIR pulses in the high intensity regime and transient X-ray absorption near-edge structure (XANES) modifications were observed.

6.1 Introduction

Extending the cut-off of the SXR source to cover the full water transparency window as discussed in Chap. 4 [1] and taking advantage of the vacuum flat jet system recently developed at ETH allowed following photo-induced chemical reactions of molecules in the liquid phase with femtosecond time resolution. These measurements demonstrate the compatibility of the developed tabletop soft X-ray source with such type of experiments. The liquid form of the sample molecules (pure or in solution) allow performing measurements under conditions closer to the natural bio environment. In addition, it allows rapid sample renewal.

In this chapter mostly the experimental measurement results and the corresponding spectral analysis are presented. The in-depth theoretical calculations of the strong-field induced dynamics were underway and are omitted from this chapter. A more detailed analysis of the results has recently been published in Smith et al. [2].

Because of its chemical composition consisting of carbon and nitrogen, the pyridine molecule C_5H_5N is a good candidate for the first time-resolved measurement in the liquid phase by looking simultaneously at two absorption edges in the water window. The C_5H_5N molecule is the simplest heterocyclic azobenzene containing

C. Schmidt, *Time-Resolved Soft X-Ray Absorption Spectroscopy of Molecules in the Gas and Liquid Phases*, Springer Theses,
https://doi.org/10.1007/978-3-030-67838-8_6

one nitrogen atom that replaces a CH group in the benzene ring. This typical ring structure can be found in a wide range of pharmaceutical compounds like vitamins B3 and B6, and many agrochemical products such as bactericide, insecticides and herbicides [3].

From a purely static point of view, a recent XANES study comparing gas to pure liquid pyridine with synchrotron using a transmission-type liquid flow cell, reported only very small energy shifts (0.1 eV) arising from the difference in the intermolecular distance of the antiparallel displaced structures [4]. The photodissociation of pyridine in the gas phase has been previously observed using translational spectroscopy [5] and multimass ion imaging [6] techniques when excited using 193 nm pulses. Furthermore, the vacuum ultraviolet 16–27 eV range induced photodissociation has been studied using photon-induced fluorescence spectroscopy (PIFS) [7] at synchrotron facilities. The photoexcitation at 193 nm revealed the elimination of a hydrogen atom directly from the closed ring together with five other two-body dissociation channels, yielding HCN, CH_3, C_2H_4, C_3H_3 and C_2H_2 fragments [6, 8]. Moreover, DFT and *ab-initio* calculations of the potential energies for many dissociation pathways have shown that isomerization of the molecule accompanied by hydrogen atom migration occurs before dissociation [6]. Besides, calculations have shown that the pyridine ring opening by C–N bond rupture is favorable because of a low activation barrier [9]. In its neutral ground state, the highest occupied molecular orbitals HOMOs of pyridine have been calculated to be the $1b_1$ (1π), $2b_1$ (2π), and $1a_2$ (3π) orbitals. The LUMOs are the $2a_2$ ($1\pi^*$) and the $3b_1$ ($2\pi^*$) [10]. The vertical ionization energies of the first two ionic states of pyridine, $^2A_1$ and $^2A_2$ have been recently measured to be 9.55 and 9.85 eV, respectively [11].

In addition to the measurements performed on C_5H_5N, this chapter also includes measurements in liquid ethanol C_2H_6O. This demonstrates the possibility of measurement at the oxygen K-edge in liquid. Due to the extremely fast dynamics induced by single strong-field ionization in ethanol, it is used as a benchmark for optimization of the experimental parameters.

6.2 Time-Resolved XANES of C_5H_5N

In order to investigate the ultrafast reaction of pyridine (C_5H_5N) under strong field ionization, the flat jet system described in Sect. 3.4.3 was coupled to the transient absorption beam line described in Chap. 2. The ionization of C_5H_5N in the liquid phase (HPLC grade) is induced by a 30 fs NIR pulse centered at 1300 nm delivering $5 \cdot 10^{13}\text{W/cm}^2$ at focus position (see Sect. 3.3.4). The lowest ionization potential of 9.6 eV has been reported by photoelectron spectroscopy studies in gaseous pyridine [12]. Considering a single ionization process and a strong-field regime with Keldysh parameter $\gamma_K \cong 0.8$, the ionization fraction of the sample is roughly estimated to be 24% using the Yudin-Ivanov model (see Sect. 2.2.5) as shown in Fig. 6.1 (left).

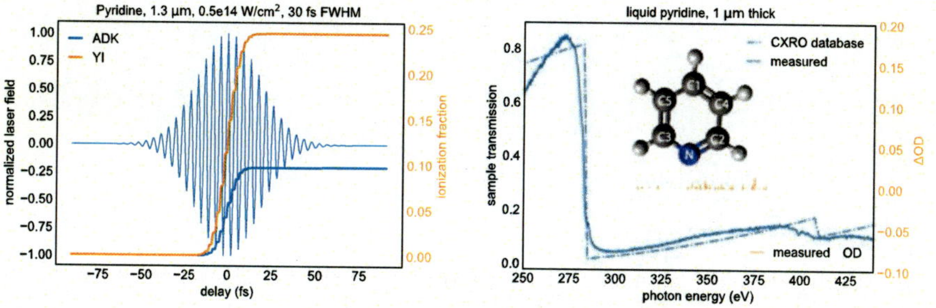

Fig. 6.1 Left: calculation of the multiphoton ionization fraction of pyridine ($IP = 9.6$ eV) using the Yudin-Ivanov non-adiabatic model for the strong field parameters of this experiment. The ADK model is also represented as comparison. Right: comparison of the measured sample transmission (solid blue) with data extracted from the CXRO database [13] for liquid C_5H_5N of 1 μm thickness (dashed blue). The orange curve shows the measured ΔOD spectrum at $\Delta t = 500$ fs. The effect of the strong laser field is visible simultaneously at the carbon (284 eV) and nitrogen (404 eV) edges. A representation of the C_5H_5N molecule is also shown

Following the transient absorption method described in Sects. 3.5.6 and 3.6, the spectra transmitted through neutral pyridine $I_0(\Delta t)$ (i.e. with pump off) and excited pyridine $I(\Delta t)$ (i.e. with pump on) are recorded at 13 delay times of 100 fs between the pump and the SXR probe pulses. In addition, the SXR spectrum without sample interaction I_{ref} is measured before and after each scan. For these measurements, the G3 configuration of the spectrometer was used (see Sect. 3.5) providing the resolutions of 0.5 and 0.9 eV at the carbon and nitrogen K-edges respectively. To prevent from transmitted and scattered infrared light of the pump beam, a 100 nm thick titanium metallic filter was installed at the entrance of the spectrometer chamber. The calibration of the spectrum is based on the carbon K-edge of C_2H_6O at 288 eV, the strong N 1s→π* transition in N_2 gas at 401 eV and the Ti $L_{2,3}$-edges at 454 eV [13] coming from the metallic filter. Figure 6.2 shows a typical scan acquired between $\Delta t = -150$ fs and $\Delta t = 1150$ fs and represented using Eqs. 3.10 and 3.11.

Figure 6.1 (right) shows the measured transmission $T = I/I_0$ through the pure pyridine liquid jet together with data extracted from the CXRO database [13] for liquid C_5H_5N of 1 μm thickness. The good agreement between the two curves confirms the estimated 1 μm of the interaction length (see Sect. 3.4.3).

From the ΔOD representation of Fig. 6.2 (left) it is possible to make the following observations. Immediately after excitation, a global increase in the absorption occurs between 280 and 288 eV and is partially separated by a local minimum around 284 eV. This energy region is located just below the previously calculated carbon atom $IP =$ 292.5 eV in pyridine [14]. The valley between the two features corresponds to the position of the main absorption peak shown in Fig. 6.2 (right) and detailed in Fig. 6.3. In Fig. 6.2 (left), the most notable transient feature is observed in the 50 to 150 fs time window around 282 eV and might correspond to an initial broadening of the main transition towards lower energies or the appearance of a new allowed transition. After 150 fs, the amplitude of this transient feature decays, leading to a narrowing of the ΔOD signal until 800 fs delay. Between 150 and 800 fs the measured signal is

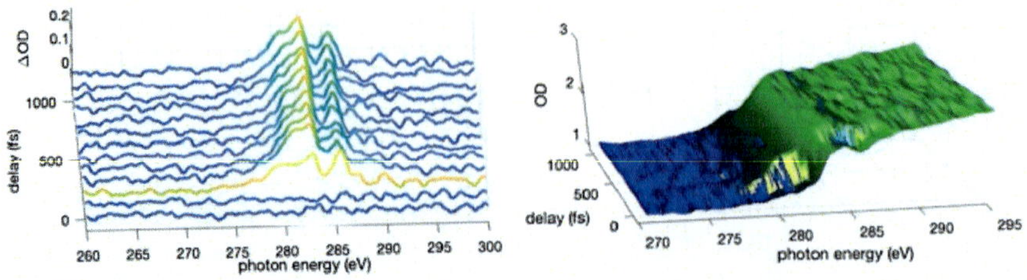

Fig. 6.2 Time-resolved XANES of liquid pyridine. Left: a waterfall representation showing the change in the XAS (ΔOD) as a function of photon energy and time delay. The positive delay times indicate that the SXR pulse follows the strong field excitation. The third time step is highlighted to emphasize the transient structure appearing just after the excitation. Right: a surface representation of the carbon K-edge absorption region showing the new feature appearing as a shoulder at the low-energy side of the main peak for positive delays. This scan was taken in 13 time steps separated by 100 fs from $\Delta t = 200\,\text{fs}$ to $\Delta t = 1100\,\text{fs}$. For each time-step, 10 measurments of both $I_0(\Delta t)$ and $I(\Delta t)$, acquired over 6 s each, are averaged together leading to a complete scan acquisition time of 30 min

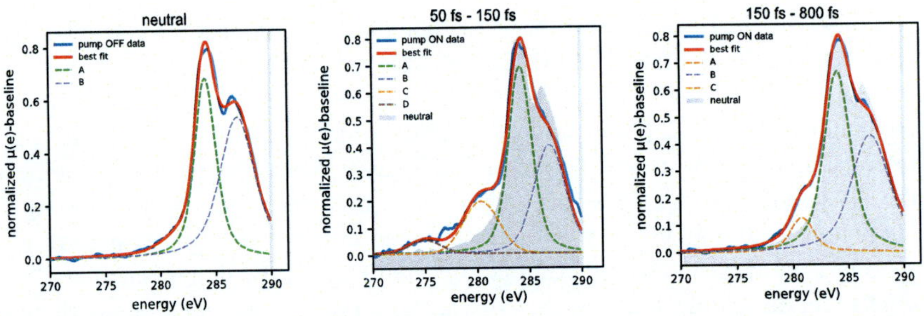

Fig. 6.3 C K-edge XANES of pyridine for 3 temporal windows of the time-resolved measurements. The strong absorption edge contributions have been removed using an identical error function centered around IP = 292.5 eV. The number of Voigt fitting functions has been chosen according to previously published DFT calculations [14] and is the minimum to correctly reproduce the spectra below the IPs. Left: spectrum of the neutral corresponding to the average over the full "pump OFF" time scan. Middle: the delayed time step highlighted in Fig. 6.2 (left) corresponding to the transient structure observed between 50 and 150 fs time delays. Right: stabilized structure of averaged "pump ON" time steps between 150 and 800 fs time delays. In order to highlight the changes in the spectra induced by the strong field interaction, the neutral spectrum is represented by the gray shaded area

almost constant which is the signature of a stabilized structure. After 800 fs and until the end of the scan, the measured signal broadens again toward the low energies. The evolution of the carbon and nitrogen XANES together with some transition assignments are detailed in the following sections.

6.2.1 Carbon 1s XANES

In order to get more insight from the measured data, the carbon XANES is extracted at two different delay windows and compared to the neutral structure. The XANES of neutral pyridine presented in Fig. 6.3 (left) is dominated by the intense band A centered at 284.5 eV and attributed to the carbon 1s→1π* transition. At higher energy, a second resolved features B centered at 286.4 eV is also visible. The energy separation between peaks A and B is 1.9 eV which promotes the assignment of the B band to 1s→2π* transition [14] or to a 1s→3sσ/$\sigma^*_{C\text{-}H}$ of final valence-Rydberg mixed orbital [10] (Table 6.1). The 1π* and 2π* orbitals derive from the 1π* degenerate orbital of benzene C_6H_6 (D_{6h} symmetry), which splits to b_1 (1π*) and a_2 (2π*) orbitals in pyridine (C_{2v} symmetry) by symmetry lowering [15]. Note that the asymmetry of the A transition, previously reported from observations and quantum calculations [4, 10, 14, 15] and attributed to the different contributions of the ortho-, para- and meta-carbon sites to the total absorption could not be resolved here due to the limited resolving power of the spectrometer as compared to synchrotron measurements.

In Fig. 6.3 (middle) representing the first step after $\Delta t = 0$ fs, the shoulders at the low-energy side of the main transition require two Voigt functions C and D to be reproduced. The observation of a similar pre-edge broadening effects occurring on a short time scale after excitation of 1,3-cyclohexadiene C_6H_8 ring structure has previously been attributed to a transient feature accounting for the 1s → 2π transition [16]. This 2π orbital being the HOMO of the C_6H_8 ring in its ground state, the short lifetime of the hole generated by photoelectron excitation indeed leads to a new transient feature on the pre-edge side. Such explanation supports the attribution of the transient structure observed just after strong field to the pyridine cation $(C_5H_5N)^+$.

The stabilized structure measured between 150 and 800 fs time delays and shown in Fig. 6.3 (right) require deeper investigations of both calculations and measurements in order to be correctly understood. Indeed, multiple direct dissociation channels together with ring opening induced dissociation channels have been previously reported from UV photoexcitation studies [3, 6] and the future calculations will allow assignment of long-wavelength ionization induced reaction pathways. The main observation is the disappearance of the D transient feature and a narrowing of the pre-edge shoulder induced by feature C. As compared to the neutral structure, the main band A also appears to be slightly broadened such that feature B at the high energy side is less resolved.

Table. 6.1 XAS transition assignment of the measured neutral C_5H_5N carbon XANES and energy comparison with previously reported XAS data

Feature	Assignment transition	This work eV	FWHM eV	Liquid [4] eV	Gas [14] eV
A	1s→1π*	284.5	2.1	285.7	285.1
B	1s→2π* or 1s→3sσ/$\sigma^*_{C\text{-}H}$	286.4	2.5	287.8	287.5

6.2.2 Nitrogen 1s XANES

From the measured time scan shown in Fig. 6.2 it is possible to observe the effect of the strong field ionization by looking at the spectrum of pyridine below the N 1s ionization threshold. As for the C 1s XANES representation, the strong absorption edge of nitrogen has been removed using an error function centered around 404.8 eV [14]. The extracted data are shown in Fig. 6.4 together with fitted absorption peaks. The neutral C_5H_5N spectrum is dominated by peak B at 399 eV ascribed to the N 1s$\rightarrow$1π* transition [4, 14]. The asymmetric line shape at the low-energy side due to the non-fully resolved structure A has been previously observed and calculated to be the signature of a vibrational fine structure in pyridine [15]. Just below the ionization threshold, the second strong feature C centered around 403.8 eV is assigned to the N 1s$\rightarrow$3π^* transition characterized by a mixture of valence and Rydberg character [15] (see Table 6.2 for comparison with previous studies). Splitting between peaks B and C is about 4.8 eV.

When looking at the stabilized structure measured between 150 and 800 fs time delays from the N XANES shown in Fig. 6.3 (right) the following observations are done. Similar to the carbon spectrum, the main feature B appears to be slightly broadened and the Voigt fitting retrieved a FWHM of 3.5 eV instead of the 2.8 eV found for the fit of the neutral spectrum. The shoulder-like feature A attributed to a vibrational structure in the neutral molecule is here redshifted and seems to broaden as well from an initial 2–3.2 eV FWHM. The C band also slightly redshifts by 1 eV

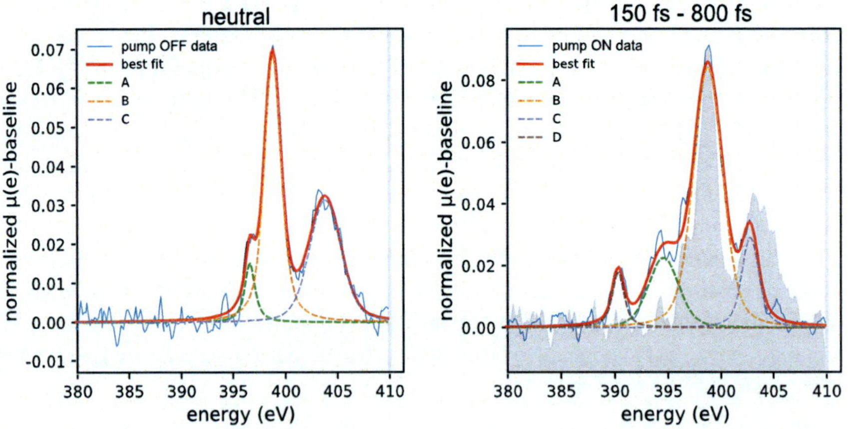

Fig. 6.4 N K-edge XANES of pyridine extracted from the time-resolved measurements of Fig. 6.1 (right). The absorption edge contributions have been removed using an identical error function centered around IP = 404.9 eV. The number of Voigt fitting functions has been chosen according to previously published DFT calculations [14] and is the minimum to correctly reproduce the spectra below the IPs. Left: spectrum of the neutral corresponding to the average over the full "pump OFF" time scan. Right: structure of averaged "pump ON" time steps between 150 and 1000 fs time delays. To directly compare the changes in the spectra induced by the strong field interaction, the neutral spectrum is represented by the gray shaded area

Table. 6.2 XAS transition assignment of the measured neutral C_5H_5N nitrogen XANES and energy comparison with previously reported XAS data

Feature	Assignment transition	This work eV	FWHM eV	Liquid [4] eV	Gas [14] eV
A		396.8	2.0	–	–
B	1s→1π*	399	2.8	398.8	398.7
C	1s→3π*	403.8	3.9	–	402.7

and tends to narrow from 3.9 to 2.8 eV FWHM. Although the most important change is the appearance of the new feature D around 390.3 eV that is not visible in the XANES of the neutral structure.

6.3 Time-Resolved Measurment of C_2H_6O

Besides the measurements of pyridine, TR-XAS of liquid ethanol C_2H_6O was also performed under similar experimental conditions. As the oxygen K-edge (540 eV) is located at higher energy than the Ti $L_{2,3}$-edges (454 eV), the titanium filter placed at the entrance of the spectrometer was replaced by an aluminum one that has high transmission in this spectral range (>80% at 540 eV for 100 nm thickness filter). Considering a single ionization process with Keldysh parameter $\gamma_K \cong 0.85$, the ionization fraction of the sample is estimated to be approximately 5% using the Yudin-Ivanov model (see Sect. 2.2.5). This lower ionization fraction, as compared to pyridine, is explained by the slightly higher ionization potential of C_2H_6O ($IP =$ 10.5 eV [17]).

The ΔOD time evolution shown in Fig. 6.5 (right) clearly indicates dynamics induced by the intense NIR pulse. The band appearing in ΔOD scan precisely at $\Delta t = 0$ fs corresponds to the instantaneous effect of ionization is stable over time. The XANES analysis exhibits no pre-edge structure and the peak visible from the

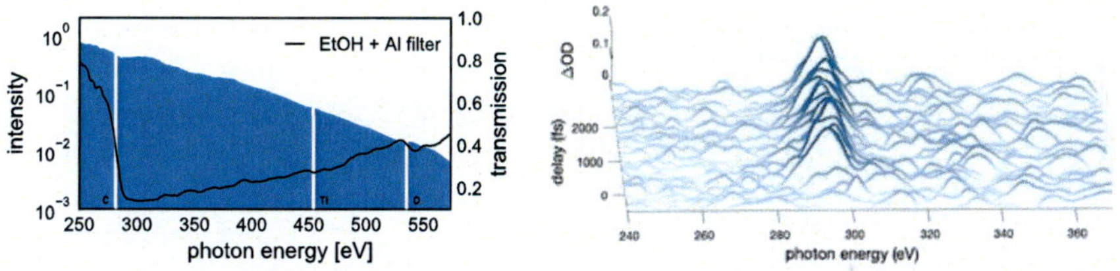

Fig. 6.5 XAS measurements of liquid C_2H_6O. Left: transmission of ethanol (EtOH) together with the SXR spectrum. The absorption step around 540 eV shows the oxygen K absorption edge in liquid ethanol. Right: a waterfall representation showing the change in the XAS (ΔOD) as a function of photon energy and time delay. The positive delay times indicate that the SXR pulse follows the strong field excitation

ΔOD representation is attributed to a global energy shift of the carbon K edge of the cation. Such interpretation is supported by *ab-initio* calculations using the Q-Chem quantum chemistry package showing only a very weak pre-edge feature at the carbon K edge appearing upon ionization. Nevertheless, these calculations predict a strong pre-edge feature induced by ionization at the oxygen K-edge with an energy separation of almost 8 eV from neutral features. Moreover, a significant change is expected in the overall oxygen XANES that should be visible in the TR-XAS measurements. This will be investigated in the future by optimizing the cutoff region of the SXR spectrum, acquiring over longer exposure times and increasing the pump intensity.

6.4 Summary

The first time-resolved measurement of strong-field induced chemical reaction in liquid pyridine has been performed by simultaneous measurement of TR-XAS at carbon and nitrogen K-absorption edges. These results demonstrate the compatibility of the developed tabletop soft X-ray source with TR-XAS measurements of molecules in the liquid phase over the water window with femtosecond time resolution.

From the observations of the first 1200 fs delay after ionization, it has been possible to identify two different molecular structures along the reaction path. The first transient structure was attributed to the pyridine cation $(C_5H_5N)^+$ which is observed during only 100 fs after the strong field pump pulse. At longer delays, the cation signature decays and a second long-living structure was observed. Advanced quantum calculations will help to correctly assign the measured electronic structure to pyridine isomers or fragments.

The XANES analysis was performed on the measured spectra by eliminating the strong edge contribution and fitting Voigt functions to the absorption bands. This analysis allowed identifying transitions corresponding to the observed structures at the carbon and nitrogen K-edges and is in good agreement with previously published synchrotron measurements.

Finally, the performed measurements in liquid ethanol C_2H_6O demonstrate the possibility of accessing the oxygen K-edge in liquid samples. This opens the possibility of studying aqueous solutions of complicated bio-compounds in the future.

References

1. Schmidt C et al (2018) High-order harmonic source spanning up to the oxygen K-edge based on filamentation pulse compression. Opt Express 26(9):11834

2. Smith AD et al (2020) Femtosecond soft-X-ray absorption spectroscopy of liquids with a water-window high-harmonic source. J Phys Chem Lett 11(6):1981–1988
3. Wasowicz TJ, Dabkowska I, Kivimaki A, Coreno M, Zubek M (2017) Elimination and migration of hydrogen in the vacuum-ultraviolet photodissociation of pyridine molecules. J Phys B At Mol Opt Phys 50(1)
4. Nagasaka M, Yuzawa H, Kosugi N (2018) Intermolecular interactions of pyridine in liquid phase and aqueous solution studied by soft X-ray absorption spectroscopy. Zeitschrift Für Phys Chemie 232(5–6):705–722
5. Prather KA, Lee YT (1994) The photodissociation of pyridine at 193 nm. Isr J Chem 34(1):43–53
6. Lin M-F et al (2006) Photodissociation dynamics of pyrimidine. J Chem Phys 124(8):084303
7. Wasowicz TJ, Kivimäki A, Coreno M, Zubek M (2014) Formation of CN (B2Σ+) radicals in the vacuum-ultraviolet photodissociation of pyridine and pyrimidine molecules. J Phys B At Mol Opt Phys 47(5)
8. Ni C-K, Tseng C-M, Lin M-F, Dyakov YA (2007) Photodissociation dynamics of small aromatic molecules studied by multimass ion imaging. J Phys Chem B 111(44):12631–12642
9. Liu R, Huang TTS, Tittle J, Xia D (2000) A theoretical investigation of the decomposition mechanism of pyridyl radicals. J Phys Chem A 104(36):8368–8374
10. Hannay C, Duflot D, Flament JP, Hubin-Franskin MJ (1999) The core excitation of pyridine and pyridazine: an electron spectroscopy and ab initio study. J Chem Phys 110(12):5600–5610
11. Śmiałek MA, MacDonald MA, Ptasińska S, Zuin L, Mason NJ (2016) Photoelectron and threshold photoelectron valence spectra of pyridine. Eur Phys J D 70(2):42
12. Kimura K, Katsumata S, Achiba Y, Yamazaki T (1981) Handbook of hei photoelectron spectra of fundamental organic compounds. Jpn Sci Soc
13. Henke BL, Gullikson EM, Davis JC (1993) X-Ray interactions: photoabsorption, scattering, transmission, and reflection at E = 50–30,000 eV, Z = 1–92. At Data Nucl Data Tables 54(2):181–342
14. Kolczewski C et al (2001) Detailed study of pyridine at the C 1s and N 1s ionization thresholds: the influence of the vibrational fine structure. J Chem Phys 115(14):6426–6437
15. Baiardi A et al (2015) Vibrationally resolved NEXAFS at C and N K-edges of pyridine, 2-fluoropyridine and 2,6-difluoropyridine: a combined experimental and theoretical assessment. J Chem Phys 143(20)
16. Attar AR et al (2017) Femtosecond x-ray spectroscopy of an electrocyclic ring-opening reaction. Science (80-.)356(6333):54–59
17. Ohno K, Imai K, Harada Y (1985) Variations in reactivity of lone-pair electrons due to intramolecular hydrogen bonding as observed by Penning ionization electron spectroscopy. J Am Chem Soc 107(26):8078–8082

Chapter 7
Concluding Remarks

Probing the ultrafast structural dynamics of matter with high selectivity and temporal resolution, be in the gas or the condensed phase, is a challenging and fascinating field of research. The access to the time scales of the early processes involving charge migration and charge transfer mechanisms, for example, offers the potential to promote optimal reaction pathways and reduce the effect of the competing ones. Such effort can be expected to lead to more selective catalysts, better energy-storage, faster data-transmission systems, and more efficient photovoltaic cells [1]. Clearly, these fundamental mechanisms can be investigated and possibly controlled only if a suitable experimental framework is developed.

This thesis work contributed to the field of ultrashort SXR pulse generation by demonstrating the feasibility of using pulse compression filamentation techniques in the SWIR regime to efficiently realize further energy upconversion by HHG. Thanks to these advancements, we succeeded in the observation of molecular photodissociation in the gas phase and field induced modifications in the liquid phase. The main results can be summarized as follows.

With the first version of the experimental setup, tabletop transient-absorption spectroscopy was demonstrated within the water window at the carbon K-edge (around 300 eV) and at the lower energy lying sulfur L-edges (160–250 eV) [2]. By achieving the highest time resolution of all X-ray studies reported to date, this work resolved and elucidated previously unobserved chemical reactions in the electronic ground states of CF_4^+ and SF_6^+ in the gas phase. The complete photodissociation process of CF_4 into $CF_3^+ + F$ has been monitored with 40 fs time resolution from the carbon 1s core electronic level. The important chemical shift of almost 10 eV induced by symmetry lowering allowed determining a maximal time duration of 152 fs until full molecular rearrangement together with the monitoring of previously reported vibrational modes in CF_3^+. Applying the same method to SF_6, confirmed the observation of Rydberg-valence mixing induced by chemical reactions and the ability of TR-XAS to monitor the evolution of unoccupied orbitals during molecular dissociation with unprecedented levels of detail.

C. Schmidt, *Time-Resolved Soft X-Ray Absorption Spectroscopy of Molecules in the Gas and Liquid Phases*, Springer Theses,
https://doi.org/10.1007/978-3-030-67838-8_7

With the objective to fully cover the water transparency window, we introduced a scheme for shortening the SWIR HHG driving pulse to less than 12 fs by using an original low-loss and efficient pulse compression procedure based on cascaded filamentation in two long gas cells. Combined with a custom differential pumping design capable of sustaining high-pressure of helium, the resulting SXR pulse enables transient-absorption measurements up to the oxygen K-shell absorption edge [3]. Moreover, the absence of waveguide in the proposed assembly considerably reduces the complexity of the optical setup as compared to equivalent state-of-the-art experiments.

Thanks to the improvement of the source, we were able to obtain a first observation of femtoseconds TR-XAS photoinduced signal from molecules in the liquid phase. Following a short NIR pulse, the XANES profile modifications were tracked simultaneously at the carbon and the nitrogen K-edges.

All of of these results are expected to have significant implications for the investigation of ultrafast structural dynamics, including processes occurring at conical intersections. The developed high-harmonic source has a range of unique advantages for time-resolved spectroscopy, for instance, enhanced signal-to-noise ratio enabled by the high repetition rate of the source, femtosecond temporal resolution and excellent temporal synchronization. Our successful demonstration of time-resolved XAS, which possesses high sensitivity to unoccupied molecular orbitals, can be a powerful complement to other techniques such as time-resolved high-harmonic spectroscopy and time-resolved photoelectron spectroscopy, which map the occupied molecular orbitals. Such tabletop experiments can be realized at the laboratory scale, which offers several advantages as compared to large-scale facilities in terms of logistics, optimization of the sample preparation procedures and long-term improvement possibilities.

In the future, the element sensitivity of XANES should be extended to the study of EXAFS with femtosecond temporal resolution that will add local geometric information in addition to the electronic structure currently monitored. With the current configuration of the setup, the measurement of biomolecules in aqueous solutions is in principle feasible. This research avenue could pave the way for studying ultrafast mechanisms in biologically relevant systems, including, for instance, thymine basis dimerization in DNA upon UV absorption, or charge transfer in porphyrins. The same is true for experiments in the solid state, which still require specific attention because of the issue of sample renewal. For instance, the investigation of weak field effects on titanium dioxide (TiO_2), might provide new insights for evaluating the transport mechanisms of charges upon light absorption in solar energy applications. Finally, the recent emergence of optical parametric chirped-pulse amplification OPCPA laser systems is presently a very promising method to efficiently generate high harmonics with longer wavelengths and reaching even higher photon energies with high repetition rate [4]. As many inner-shell absorption edges of catalytic, magnetic and strongly correlated electron systems (Fe, Co, Ni, Cu) lie at photon energies up to the keV, accessing this regime will create breakthrough opportunities in the use of light sources such as the one developed during this thesis.

References

1. Wörner HJ et al (2017) Charge migration and charge transfer in molecular systems. Struct Dyn 4(6):61508
2. Pertot Y et al (2017) Time-resolved x-ray absorption spectroscopy with a water window high-harmonic source. Science (80-) 355(6322):264–267
3. Schmidt C et al (2018) High-order harmonic source spanning up to the oxygen K-edge based on filamentation pulse compression. Opt Express 26(9):11834
4. Popmintchev T et al (2012) Bright coherent ultrahigh harmonics in the keV X-ray regime from mid-infrared femtosecond lasers. Science (80-) 336(6086):1287–1291

Appendix
Table of X-Ray Excitation Energies of the Elements

See Fig. A.1.

C. Schmidt, *Time-Resolved Soft X-Ray Absorption Spectroscopy of Molecules in the Gas and Liquid Phases*, Springer Theses,
https://doi.org/10.1007/978-3-030-67838-8

X-ray Excitation Energies of the Elements

Symbol	Z
Symbol Z →	Cu 29
K →	8.979
L_I →	1.100
L_{II} →	0.952
L_{III} →	0.932

Energies are in KeV

H 1 0.014																	He 2 0.025
Li 3 0.055	Be 4 0.112											B 5 0.188	C 6 0.284	N 7 0.402	O 8 0.537	F 9 0.686	Ne 10 0.867
Na 11 1.072	Mg 12 1.305 0.063 0.050 0.049											Al 13 1.560 0.087 0.073 0.072	Si 14 1.839 0.118 0.100 0.099	P 15 2.149 0.153 0.136 0.135	S 16 2.472 0.193 0.164 0.162	Cl 17 2.822 0.238 0.202 0.200	Ar 18 3.202 0.287 0.251 0.248
K 19 3.607 0.341 0.297 0.295	Ca 20 4.038 0.400 0.350 0.346	Sc 21 4.493 0.463 0.404 0.399	Ti 22 4.965 0.531 0.461 0.454	V 23 5.465 0.604 0.520 0.512	Cr 24 5.989 0.682 0.584 0.574	Mn 25 6.540 0.754 0.650 0.639	Fe 26 7.112 0.842 0.720 0.707	Co 27 7.709 0.929 0.793 0.778	Ni 28 8.333 1.012 0.872 0.855	Cu 29 8.979 1.100 0.952 0.932	Zn 30 9.659 1.196 1.044 1.021	Ga 31 10.367 1.302 1.142 1.115	Ge 32 11.104 1.414 1.249 1.218	As 33 11.868 1.530 1.360 1.325	Se 34 12.658 1.653 1.477 1.436	Br 35 13.474 1.782 1.596 1.550	Kr 36 14.322 1.920 1.726 1.675
Rb 37 15.200 2.065 1.863 1.805	Sr 38 16.105 2.216 2.007 1.940	Y 39 17.080 2.373 2.156 2.080	Zr 40 17.998 2.532 2.307 2.223	Nb 41 18.986 2.698 2.465 2.371	Mo 42 19.999 2.866 2.625 2.520	Tc 43 21.045 3.043 2.793 2.677	Ru 44 22.117 3.224 2.967 2.838	Rh 45 23.220 3.412 3.146 3.003	Pd 46 24.350 3.605 3.330 3.173	Ag 47 25.514 3.806 3.524 3.351	Cd 48 26.711 4.018 3.727 3.537	In 49 27.940 4.238 3.938 3.730	Sn 50 29.200 4.465 4.156 3.929	Sb 51 30.491 4.698 4.381 4.132	Te 52 31.813 4.939 4.612 4.341	I 53 33.169 5.188 4.852 4.557	Xe 54 34.582 5.452 5.100 4.781
Cs 55 35.985 5.713 5.359 5.012	Ba 56 37.441 5.987 5.624 5.247	La 57 38.925 6.267 5.891 5.483	Hf 72 65.351 11.272 10.739 9.561	Ta 73 67.414 11.680 11.136 9.881	W 74 69.524 12.098 11.542 10.204	Re 75 71.676 12.525 11.957 10.534	Os 76 73.872 12.964 12.384 10.871	Ir 77 76.112 13.424 12.824 11.215	Pt 78 78.395 13.892 13.273 11.564	Au 79 80.723 14.353 13.733 11.918	Hg 80 83.103 14.846 14.209 12.284	Tl 81 85.528 15.344 14.698 12.657	Pb 82 88.006 15.860 15.198 13.035	Bi 83 90.527 16.385 15.708 13.418	Po 84	At 85	Rn 86 98.417 18.055 17.334 14.612
Fr 87	Ra 88	Ac 89															

Ce 58 40.444 6.549 6.165 5.724	Pr 59 41.991 6.835 6.441 5.965	Nd 60 43.569 7.126 6.722 6.208	Pm 61 45.184 7.428 7.013 6.460	Sm 62 46.835 7.737 7.312 6.717	Eu 63 48.520 8.052 7.618 6.977	Gd 64 50.240 8.376 7.931 7.243	Tb 65 51.996 8.708 8.252 7.515	Dy 66 53.789 9.047 8.581 7.790	Ho 67 55.618 9.395 8.919 8.071	Er 68 57.486 9.752 9.265 8.358	Tm 69 59.390 10.116 9.618 8.648	Yb 70 61.332 10.488 9.978 8.943	Lu 71 63.314 10.870 10.349 9.244
Th 90 109.649 20.470 19.692 16.300	Pa 91	U 92 115.603 21.756 20.947 17.167	Np 93	Pu 94 121.760 23.095 22.263 18.053	Am 95	Cm 96	Bk 97	Cf 98	Es 99	Fm 100	Md 101	No 102	Lr 103

Fig. A.1 Table of X-ray excitation energies of the elements. The transitions lying in the 150–540 eV energy range are highlighted in yellow. Data from [1], adapted from: M. Newville

Reference

1. McMaster WH, Kerr Del Grande N, Mallett JH, Hubbell JH (1970) Compilation of X-ray cross sections UCRL-50174, sections I, II revision 1, III, IV. At Data Nucl Data Tables 8(4–6):443–444

C. Schmidt, *Time-Resolved Soft X-Ray Absorption Spectroscopy of Molecules in the Gas and Liquid Phases*, Springer Theses,
https://doi.org/10.1007/978-3-030-67838-8